The Book of

Rubbish

Ideas

An interactive room-by-room guide
to reducing household waste

Tracey Smith

This book is dedicated to Alex, Ryan and Abby
And your children
And your grandchildren
The leaders of the future

A donation from the author's royalty will be made to NACOA
(National Association for Children of Alcoholics)

Design and Typesetting: Bookcraft Ltd
Illustrator: Felix Bennett (with thanks to Dover Publications)
Project Manager: Lyn Hemming
Production: Anny Mortada

First edition of The Book of Rubbish Ideas
Copyright © September 2008 Alastair Sawday Publishing Co. Ltd
Fragile Earth – an imprint of Alastair Sawday Publishing Co. Ltd
Text © Tracey Smith

Published in 2008
Alastair Sawday Publishing Co. Ltd
The Old Farmyard, Yanley Lane,
Long Ashton, Bristol BS41 9LR
Tel: +44 (0)1275 395430
Fax: +44 (0)1275 393388
Email: info@sawdays.co.uk or info@fragile-earth.com
Web: www.sawdays.co.uk or www.fragile-earth.com

The publishers have made every effort to ensure the accuracy of the information in this book at
the time of going to press. However, they cannot accept any responsibility for any loss, injury or
inconvenience resulting from the use of information contained therein.

ISBN-13: 978-1-906136-13-0

Printed in Cornwall by TJ International Ltd on FSC-certified paper using soya-based inks.

Contents

Foreword

As this century unfolds, our children face more changes and uncertainty in their everyday lives than at any other time in history.

It's difficult for us to step back and imagine what 'their' world would be like without mobile phones, plasma television or even the humble electric light, and inconceivable for the children of the new millennium to contemplate a life without today's electrical techno-gadgetry.

Many advances were made in the last century that have allowed us to live our lives 'conveniently' and with no care or consideration for where things are sourced or where they'll eventually end up; so who could blame our children for having difficulty re-connecting with the 'make do and mend' values of our grandparents?

Thankfully, the tide is turning! In the last few years the word 'provenance' has found a more common use in everyday conversation. The 'Bra Wars' in 2007 prompted people to think about where and how our clothes are made, whilst the high-profile campaigns led by chefs Jamie Oliver and Hugh Fearnley-Whittingstall have encouraged us to question where the chicken we eat comes from.

Plastic bags are on their way out, seasonal local produce is in, high food miles are unfashionable and discussions about the environmental legacies we are leaving for future generations are commonplace. All this eco-talk can feel overwhelming and it is difficult to believe that one person changing a few disposable habits can make a difference, but it can and will.

It has never been more important to spread encouraging and positive messages about sustainable living and this wonderfully accessible book will enable you to be part of the green revolution that's taking place.

Brigit Strawbridge

Writer, Broadcaster & Creator of The Big Green Idea
www.thebiggreenidea.org

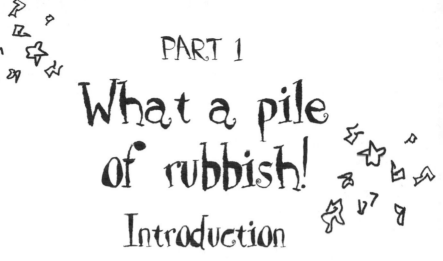

PART 1

What a pile of rubbish!

Introduction

As you are reading this book, I am assuming that you're up for the idea of embracing a little change in the way you handle your rubbish. The book will introduce you to ways in which you can make a difference in your everyday life.

In Part 1

we talk about the serious side of rubbish. Rubbish is a serious business, and recent figures show that, in England alone, we generate a shocking 100 million tonnes of waste per annum from households, commerce and industry. I've barely started and I already feel the need to make a confession. When I was first asked to write this book, I thought it was going to be easy. My brain overflows with simple

ideas on how to cut out the excess packaging we are surrounded by, on de-cluttering homes, on eradicating waste and rubbish in general and, while doing all of the above, saving my hard-earned cash and being kinder to my local and global environments. However, the deeper I plunged into my research, the darker and more cavernous the whole hole of knowledge became. I almost fell into a well of statistics that I needed a rope to hoist me out of. Therefore, while I promise to keep the 'serious bit' to a minimum, it's definitely staying in. Don't worry, you can always flick forward and lose yourself in a bit of idle distraction by colouring in one of the eco-cartoons.

In Part 2 we really get to grips with it all. I will be tiptoeing through your home, room by room.

I won't look in your drawers, but I probably won't be able to fight off the urge to poke about under the bed. I will focus on the key areas where we have excessive rubbish, or experience unnecessary waste in the way of packaging, resources, utilities, money, time, travel, shopping, storage and so on. There are some instant solutions to these situations, many that won't cost you a penny and others that require varying degrees of financial investment, but nothing that's going to break the bank.

This section also includes my own tried, tested, inexpensive and eco-friendly recipes for household cleaning. But don't panic, there is absolutely no chance of me banging on about housework, oh no. But I will be showing you realistic ways to stop buying copious amounts of those toxic, plastic-bottled and tinned sanitising products that we seem to have been seduced into buying over the past few decades. There is also a toolbox of letters interspersed throughout the rooms for you to get active with, and get your friends, local shops and school behind the drive for waste reduction.

Part 3 is a rubbish resources round-up. There is information on useful websites, further reading material,

addresses and contact details of great organisations and some very helpful people I've met along the way. For anyone who enjoys a bit of celebrity gossip, there's even a bit of that too. I'll be exposing who's green and means it, who's doing what, with whom, how they are doing it and why they're bothered. The life of a celebrity is mightily powerful, and if Jennifer Aniston can influence the trend of a nation with a haircut, then I might stand a chance of getting you to change your ways.

Naturally, this will remain a work in progress beyond the publication of this book, but I'm in this for the long haul and I'll be reporting 'hot and happening' eco-news on a dedicated blog (**www.bookofrubbishideas. co.uk)**, where you can see what's going on and further your progress to reduce your rubbish. You can also submit your questions for the celebrities, for 'my people' to submit to 'their people'. You never know, we might even get replies. You can ask general questions too, that I hope I can answer for you.

Collectively, the information in this book should enable you to live more simply, with dramatically reduced levels of rubbish. So before we go any further, let's make a pact.

I promise to give you all the tools you need to:

🌸 clear out the superfluous stuff that's cluttering up your home and your life

🌸 learn how to avoid unnecessary rubbish and waste

🌸 help you see 'rubbish' as a resource

🌸 save rubbish, time, money and more by doing all of the above.

When you get to the last page, you shouldn't have a burning desire to pack it all in and go live in a yurt ... (well, not unless you were planning to do that anyway, in which case don't fight it; get packing).

....

Now it's your turn.
You promise to:

    🌱 read this book to the end (not all in one go of course), paying attention to the bits you would usually skip through

    🌱 take at least one tip from each chapter and implement it, ideally later that day

    🌱 make a nice cup of tea (Fairtrade of course; or coffee, or hot chocolate), set aside ten minutes of time, hop onto the blog and browse through my top tips and movies, showing you rubbish reduction in action.

When you've finished with this book, pass it on to a good friend who could do with finding a greener groove.

....
(Please sign and date here, then crack on and get reading!)

....
(Signature space for subsequent readers)

....

And don't forget — landfill is only one small step away from landfull ...

Rubbish history

It's bin day, the bag's not out, and I've only got fifteen minutes to walk the kids to school. But hang on, which bin day is it? Is it 'recycling' day or 'normal dustbin' day? A quick look down the street will soon tell me. We don't have wheelie bins, and I can see an array of black bags and various plastic carrier bags, mainly from supermarkets, duly 'recycled'. Some of these were clearly put out last night, and the local cats, birds and other wildlife have had a rummage through for breakfast.

Anyway, my question's answered, it's 'normal dustbin' day. Panic over. I send one of the children in to grab our bag and leave it by the gate.

On average, each person in the UK throws away his or her own body weight in rubbish every seven weeks.

Had it been recycling day, it would have been a different matter. I take my recycling seriously and not much finds its way into the 'black bag', but I'm not perfect by a long chalk, and every time I flip the lid up on my kitchen bin it narks me.

I'm a writer on sustainable living, and I *know* all about the 'right' things to buy when I go shopping; no fruit and vegetables in unnecessary containers or wrapping; only fresh cuts of meat that are sold 'loose' from the butchers; and definitely no plastic bags as a matter of principle. Yet still I have a little black bag and, regardless of its small size, it does exist and continues to fuel a raging fire in the soul of my enthusiasm to reduce it even further next week and maybe, some day soon, to eliminate it completely.

Recycling was probably taking place in various forms before the written word recorded it. One person's rubbish is another's treasure and, rather unsurprisingly, the Greeks and Romans had reasonably efficient structures in place early on to deal with their rubbish. One of the first documented municipal landfill sites was in Crete in 3000BC, where it was noted that all manner of rubbish and waste was dumped into enormous pits, dutifully covered in earth and left for archaeologists to ferret around in hundreds of years later. In 500BC, a law was passed in Athens that stated that all rubbish had to be deposited at least a mile from the city walls at an organised site; there were sections where you could also leave organic waste to compost. This ancient, well-organised civilisation was not new to innovation, and regularly re-used and recycled many materials.

The simplest solutions are usually the best, and the Romans' approach to rubbish could have set a precedent for other nations, but history tells another story. By the 11th century, many of the sensible rubbish disposal and recycling ideas had simply been forgotten in this country. As time went on, rubbish was either burnt or, more often, discarded at the point of origin, and there were piles of what can only be described as utter filth on the streets.

This was made worse by almost non-existent options for ridding oneself of daily effluent, and many took to tipping it directly into the gutter or their own cesspit, which cost Londoners a shilling to empty; others, who clearly couldn't stand the stench, risked fines and prosecution by tipping it discreetly into the nearest waterway. Many neighbourhoods kept pigs to deal with the organic waste that was created; the pigs would be eaten at some point in the future. Dogs and assorted vermin took care of any remaining organic waste. In the mid-1300s, flea-infested rats introduced the bubonic plague or 'Black Death' to our shores. The rats thrived in the putrid and squalid conditions, and over a three-year period the Grim Reaper delivered suffering and a hideous death to about a third of the population of Europe.

The first Public Health Act was introduced in 1848, the aim being to improve the sanitary conditions of towns and populous places in England and Wales. It legislated placing the supply of water, sewerage, drainage, cleansing and paving under a single local body. These 'bodies' were required to clean the streets in each district, removing dust, ashes, rubbish, filth, dung, soil, obstructions and nuisances. So began a slow migration towards a more sanitised nation. By the mid-1800s, the population was on the rise again, and London was officially the most polluted city in the world. The River Thames, which snaked its way through the city, was so contaminated by effluent and rubbish that 1858 was officially called 'The Year of the Big Stink'.

An industrious group of individuals realised there was money to be made in the collection and disposal of rubbish, and began to collect ash from the grates of coal-burning fires, thus earning the fitting title of 'dust' men and women. They sold this precious commodity on to brick-makers who used it in the production thereof, and also to anyone that needed a soil conditioner or fertiliser. They also filtered out and sold on other objects that made their way onto the carts, such as nails, bones and rags. This was the origin of our homegrown recycling.

Thankfully we've come a long way since then, although some would argue we've not come anywhere near far enough. As a nation, we've fought many different battles since the plague but, as a planet, perhaps we face a far bigger danger with the problems we're creating for ourselves as a result of super-seduction by mass-consumerism. We are living in what might historically be remembered as the 'take-away, throwaway era'. The drive to earn more and possess more is ever-present. Our children are growing up with the mantra that it's 'cheaper' to replace than repair. We love an instant fix to our problems. Almost everything has a remote control; there are thermostats, instant touch controls and timer settings for heating, lighting, listening, cooking, and our busy lives dictate a craving for ready-made, pre-packed solutions. Our insatiable need for a multitude of products that, allegedly, make our life simpler and more convenient, is being met by manufacturers and food producers alike, and prices are artificially being driven down to keep a constant flow of sales.

Advances in technology continue to provide manufacturers with thinner and more efficient types of packaging and wrapping materials, which extend the shelf lives of their wares. But the problem of disposal of all of this excessive 'waste' material is left at the door of the consumer, which nudges our ever-increasing population a step closer towards a death by plastic.

Another area of concern is the alarming amount of disposable products that are ingrained in our daily life — the ballpoint pen on your desk, the lighter in your pocket and the razor that you used this morning. We need to find ways to reduce this type of waste and to recycle what we can't eliminate. Landfill and incineration are at the moment the two most commonly chosen methods of disposal, but both come with an unacceptable environmental cost.

There are simply no acceptable excuses as to why organic waste should find its way to landfill. Home and community composting offers an immediate solution; there are an increasing number of food and garden waste collection services springing up all over the country, and you can usually deposit green waste at your local municipal site too.

When organic materials, such as food and garden waste, are deprived of oxygen they begin to decompose anaerobically. This anaerobic biodegrading produces methane, one of the most potent greenhouse gases. Landfill sites are packed with assorted sealed rubbish bags containing organic waste, and are now a prime source of methane pollution.

The EU Landfill Directive states that by 2020 we must reduce all biodegradable municipal waste heading to landfill, which includes the contributions from households, to 35% of what it was in 1995. Before then, however, we have to reduce the amount to 75% of the 1995 levels by 2010 and then to 50% of the 1995 levels by 2013.

This would be a magnificent achievement, but I fear a great deal more will have to be done to support waste reduction campaigns such as WRAP and Recycle Now, to help them and other initiatives enthuse the general public to get them more involved. For example, it is paramount that collection of organic waste be increased to include all homes and businesses; that more consistent, countrywide incentives be provided, such as reduced-cost composting bins from local authorities for personal and community composting schemes; and that legislation be put in place to ensure all new building developments include composting facilities. We must start accepting personal responsibility for our organic waste. By bringing sustainable-living lessons into schools by way of the national curriculum, we could raise awareness and create a great foundation for the leaders of the future.

As for disposal of the rest of our rubbish, we are presented with the choice of the two equally unattractive and very ugly sisters — Landfill and Incinerator.

News stories about landfill and the fact that we are rapidly running out of room and unable to continue with this method of disposal sustainably seem to be hitting the headlines with ever increasing frequency, but we hear less about incineration. It only makes the headlines when proposals to build a new site are submitted, and local residents stand up to be counted and have their worried voices heard.

Old-style incinerators are inefficient and often have no way of separating materials, of removing items that *could* be recycled before processing or of extracting any hazardous compounds. The health of local residents and, of course, plant workers is potentially at risk. To add insult to possible injury, many don't even attempt to harness the energy generated through burning as a source of heat and potential power.

In an attempt to make incinerators more appealing, new and more efficient versions have been designed. One of the new terms for the modernised facilities that harness power created is 'Waste to Energy Plants'. This has been coined to help this procedure sit more comfortably in the government's waste management strategy. However, it does little to appease the growing number of anti-incinerator groups that are forming all over the country, some uniting with other citizens worldwide to share their concerns for the potentially harmful short- and long-term effects on personal health and on the immediate and global environment. And incineration does little to encourage waste reduction. Often labelled as recycling facilities, converting waste into energy, they are driving the demand for waste to feed their fires. There is no consideration of the inefficiency of creating energy this way, burning materials that could be more valuably recycled in more conventional ways. Or of the vast amount of emissions given off by incineration, which adds significantly to the greenhouse effect and climate change at a time when the global population is searching for ways to decrease these gases.

Harnessing heat and power from this method of waste disposal has become the forte of countries like Denmark and Sweden, who are leaders in the field with super-efficient systems. It is also a popular choice in Japan where available land is scarce.

In the UK, generally, the technology employed for the more efficient style of incineration seriously lags behind that of our European neighbours, as historically we've appeared complacent and, clearly, reliant upon the options presented by landfill. However, in an attempt to drastically reduce the amount of greenhouse gases produced by our landfill sites, The Landfill Directive, a piece of legislation created by the European Union, has eventually led our government to impose new landfill taxes and other initiatives such as the landfill allowance trading scheme. These sit alongside a series of other measures designed to reduce the rubbish sent to landfill and encourage a greater take-up of recycling. However, some argue that simply by promoting incineration as a viable option, we negate the increasing efforts that present recycling and composting as the only clear long-term winners.

For now though, it seems that burning the contents of our black bags will continue to play a part in the future strategy of dealing with municipal waste, and with oil now exceeding $143 a barrel it seems to hold appeal as an alternative energy source, despite evident public reticence and questionable efficiency.

At a community level, it's clear to see that the points against mainstream adoption of incineration far outweigh the points in favour of it. This comes despite the claims that high-temperature treatments of material such as toxic medical waste will *apparently* render many of them sterile and harmless. People have serious health concerns about the chemicals that are being released into the atmosphere from incineration.

Of concern to lobby groups that oppose incineration are nitrogen oxides, sulphur dioxide, hydrogen chloride, carbon dioxide and two of the worst offenders — dioxin and furan. Incineration is believed to be the main route by which dioxins and furans are produced and is often the area of focus in pollution-prevention efforts. Scientific studies have shown that they can produce a range of effects on animals and humans, including skin disorders, liver problems, impairment of the immune and endocrine system and reproductive functions, effects on the developing nervous system, and certain types of cancers.

Of course, other lifestyle factors can exacerbate exposure to these toxins, but the fact remains that they *are* a by-product of burning, and damage limitation seems to me a logical way forward.

The future feasibility of this process is uncertain given that more paper and other combustible materials are now being recycled than ever before. There are concerns that other fuels (e.g. natural gas) may be required to aid the process of burning! Finally, we are still faced with the problem of what to do with the considerable amount of ash produced by the furnaces; I fear it may still find landfill as its final resting place.

Every step we take towards reducing our rubbish, no matter how small, is important. It's unhelpful to offload guilt on others, but busy parents especially often set themselves up as easy targets when it comes to eco-sins. To parents, disposable is often seen as 'the norm'. I remember the storeroom in the maternity unit when I had my three children over ten years ago. There were shelves of disposable nappies, stacked alongside a mountainous pile of single-use changing mats and boxes of 'Baby Packs' full of disposable wipes, throwaway breast pads, and freebies to take home; there wasn't a washable, cotton item to be seen!

Disposable nappies alone account for between 400,000 and 500,000 tonnes of waste in landfill sites across the UK. Organisations such as The Real Nappy Campaign are devoted to finding ways of encouraging parents to try washable options, but if the idea of using a washable nappy hasn't been 'normalised', new parents are less likely to try them. Perhaps the media should work harder at taking the 'freaky out of eco' for the benefit of the greater good. There have been some incredible advances in nappy design and functionality, and the leaky terylene, wrapped up like a bad towel and fastened with a fat safety-capped pin, is a thing of the past. My children have been a long time out of nappies and I didn't take the cotton washable option, which leaves me wracked with green guilt.

The magazine rack of your local newsagent will reveal a plethora of holistic magazines that cater for those interested in organic, fresh food cooking, growing their own vegetables, making their own yoghurt, and many other 'green' ways to live. However, there is little useful information for the regular person who needs a little help and encouragement to get started.

This book will help. We must build upon our successes, look at the progress that we've made so far, and start to believe that our small, collective changes really *can* make a difference.

Consider the plastic-wrapped fruit and vegetables. If we stopped buying them in plastic, the shops and supermarkets might eventually stop stocking them in this manner. To enforce permanent changes of this magnitude would involve legislation, planning, discussion, letter-writing, meetings, coffee-drinking, contingency plans and diagram-drawing. Perhaps a simpler option would be for customers to ditch excess packaging at the till for the shopkeeper to recycle responsibly. This would be a good first step forward.

Household waste is composed of a wide variety of materials. The best overall estimates currently available show that household waste typically consists of garden waste (20% of the total), paper and board (18%), putrescible waste such as kitchen waste (17%), general household sweepings (9%), glass (7%), wood and furniture (5%), scrap metal/white goods (5%), soil (3%), textiles (3%), metal packaging (3%) and disposable nappies (2%).

Individuals cannot shoulder all the blame. For every tonne of household waste produced, commercial, industrial and construction businesses produce another six tonnes. Fortunately, awareness of corporate social responsibility and new government initiatives, are helping to quash, or at least reduce these levels at source.

There's a lot of groundwork to be done, with basic re-education for all members of society, of all ages, on how we can better dispose of our rubbish. The solutions lie in our hands. We have the power to effect change on a local and global scale, but before we set out to tackle the conglomerates, we would do well to make a few fundamental changes at home first. Perhaps my ambition to cure the world's ills in one fell swoop is a little optimistic, and my focus must remain steadfastly on evoking willing change at an individual level.

People live in homes, they go to work and to places of learning, and all three of these venues would benefit from simple rubbish and waste reduction notions being implemented by enthused individuals, so let's strip it all right back to you and me and get started!

Rubbish climate

Throughout 2007 we watched weather-related disasters play out beyond our living room windows and on our television screens. The merciless winds of El Niño wreaked havoc in southern Europe, and there were freakish conditions of hot, cold and wet, flooding and landslides in the UK. In May 2008 a tropical cyclone in Burma claimed over 100,000 lives and left millions without food, water, shelter and medical care. Many environmental experts in search of an explanation for these extreme conditions have linked the events to the inevitable effects of climate change.

Of course, for every expert there are at least two debunkers, but the startling fact remains that nine of the ten warmest years on record have occurred since 1990 and climate change has been labelled as the probable cause. Matthew Oates, the National Trust's nature conservation adviser in the UK, commented that "2007 was an utterly unique year, full of extremes. This summer was a wake-up call for those who don't believe in the actuality of radical climate change". During 2007, comments such as this were made in the news almost daily, many from respected authorities on climate and environment. One positive outcome has been the deluge of suggestions on ways we can all 'do our bit' to prevent climate chaos from getting any worse. This is backed up by the *Stern Review*, which shows that the cost to the economy of tackling the threat of climate change now will be far less than the damaging costs of the consequences later if we fail to take prompt action (*The Economics of Climate Change: The Stern Review*, Nicholas Stern for HM Treasury, 2006).

There is general agreement that one way forward is for us all to reduce our physical impact on the environment; minimising our household rubbish is an easy place to start. To understand the impact that our rubbish has, we need to know what happens to it, recycling and otherwise, once we have liberated it from our homes. DEFRA (the Department for Environment, Food and Rural Affairs) is responsible for national legislation on waste throughout England. They have set out a

policy framework for all local authorities to work from
and a helpful, if wordy, public report called 'Waste Strategy for
England 2007' can be found on their comprehensive website. The
document sets out their vision for sustainable waste management
for some years to come. Each individual local authority decides what
to do in its area and acts accordingly. It's a truly complex structure.
In general, the district or smaller councils are responsible for the
collection of household rubbish, but the larger shires and county
councils are responsible for its ultimate disposal. There are 121 waste
disposal authorities in England alone, most singing from their own
hymn sheet.

With a layman's eye, it might seem a logical solution for these
authorities to work together, share their resources and for all councils
to do the same thing. I live on the edge of three counties; all play the
same game but by completely different rules. A friend further along the
road can put out food waste with her rubbish and it will be taken to an
anaerobic digester — a very big composting bin that works at a higher
temperature than a home compost bin. But I can't!

However, DEFRA does not support a unified system for collection
and disposal, simply because every village, town, city and hamlet is so
different. They do urge neighbouring authorities to work together where
appropriate, but cannot enforce this. With sharp increases in landfill
taxes planned for the future it is clearly in the authorities' financial
interest to make the numbers work.

The aim of the directive is to reduce global greenhouse gas
emissions from the waste sector by 9.3 million tonnes of carbon
dioxide equivalent per annum. Put in simpler terms, that's roughly
the same amount as is produced by three million cars on the road.
To nudge things along with the aid of a stick and carrot, the report
also contains Landfill Directive Diversion Targets for 2010, 2013
and 2020, and a spokesman indicated that 'we' (England that is) are
making good progress towards achieving the first of them; recycling
and composting of waste increased by 27% in 2005/6; recycling of
packaging waste has increased by 56% since 1998, and between

2000/01 and 2004/5 there has been a 9% reduction of waste being sent to landfill, with municipal waste growing at the rate of just 0.5% per year.

That said, we are still dragging our heels in comparison with the progress of many of our European neighbours, and the figures of total municipal waste are still alarming. To inspire change within the local authorities themselves is a Herculean challenge, but to get the public on board with the idea of necessary waste reduction is surely the greatest challenge of all. The simple fact remains: with the exception of a very few authorities who burn some household rubbish to create fuel, almost all black bag and bulging wheelie bins collected on 'normal rubbish' day end up sitting amongst a stinking pile of other bags in the middle of a vast site we know as landfill.

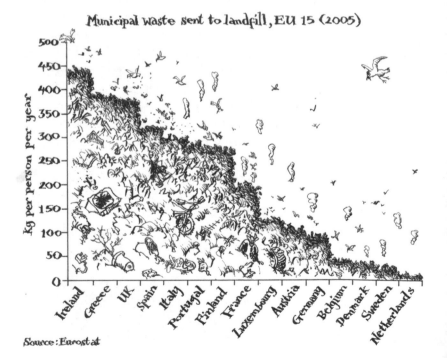

Municipal waste sent to landfill, EU 15 (2005)

kg per person per year

Ireland, Greece, UK, Spain, Italy, Portugal, Finland, France, Luxembourg, Austria, Germany, Belgium, Denmark, Sweden, Netherlands

Source: Eurostat

The news about recycling, however, is a little more uplifting. Recyclate, the general term for all manner of recyclable materials, is sorted and sold on to companies who turn the resource into new products. Re-using recyclate is usually more economical in financial terms, and definitely better in sustainability terms, than sourcing the raw materials to make these products from scratch. The easily recognisable and well-used sign depicting three arrows set in a triangle – indicating 'recycle' – means that particular item has been made using recycled materials and/or that it is suitable for recycling.

In 2006/07 local authorities reported that around eight million tonnes of household waste (31% of total household waste) were diverted for recycling or composting through schemes run by local authorities or organisations working in partnership with them. In 2006/07 the largest component was compost (36%); this was followed by paper and card (19%) and then co-mingled materials (15%). Co-mingled collections – the collection of a number of recyclable materials in the same box or bin, e.g. paper, cans, plastics – have become more widespread in recent years. A material breakdown of household recycling over time, and by region, is available from DEFRA (see Resources).

How long does your rubbish take to break down?

Banana peel	2–10 days
Cotton rags	1–5 months
Sugarcane pulp products	30–60 days
Paper	2–5 months
Rope	3–14 months
Orange peel	6 months
Wool socks	1–5 years
Cigarette filters	1–12 years
Leather shoes	60–80 years
Nylon fabric	100 years +
Plastic six-pack holder	450 years
Nappies	300–500 years
Aluminium cans	200–400 years
Plastic bottles	450 years – never
Chewing gum	never
Car tyre	1,000s of years – never

NB: Most of the times are speculative and many of the items listed will never biodegrade, they will only fragment into smaller and smaller pieces over time.

Most of us are familiar with commonly recycled items such as aluminium, tins, glass, paper, card, food, kitchen organic waste and garden waste. More obscure items include books, batteries (household and car), oils (cooking and car), textiles, pots of paint, and even electrical appliances and mobile phones. The disposal of plastic continues to pose a great problem.

What exactly happens to rubbish at the sorting centre? First, everything has to be sorted and cleaned. This is often a manually assisted process, and contaminated items, such as half-empty cat food tins, run the risk of being completely rejected, so rinse them out!
• **Aluminium cans** are shredded, melted down, and the molten aluminium is poured into giant ingot moulds. The ingots are sold to make

new products such as car and aeroplane parts and, of course, new cans for the food and drink industry.

• **Steel** is a brilliant product to recycle, as it can be reprocessed again and again. Steel cans are melted down in a furnace and combined with other raw materials such as molten iron. The hot steel is then cast into solid slabs and rolled into foil to make new cans. Some steel products, e.g. stainless steel, are made from 100% recycled material.

• **Glass** is one of the most versatile recycled products. It can be crushed to make a substance called cullet and added to the mix of raw materials — sand, soda ash, limestone and other additives for colouring or special treatments — that make up new glass containers. Recycled glass is also used to make less commonly known items, e.g. bricks for construction, and is turned into 'Glassphalt' for road surfacing.

Fourteen million bottles were crushed and used to resurface the M6 motorway.

Glass is also used as a filtration medium for swimming pools and, even more bizarrely, can be turned into fine sand for golf course bunkers.

• **Paper** is pulped, cleaned and then screened. The ink on printed paper from newspapers and magazines is removed with soap and bubbles of air. The pulp is then turned back into paper via a huge mechanised system where it's drained, pressed, dried and then made into enormous reels or sheets.

• **Cardboard** is initially recovered, bulked and sold to paper and cardboard mills. It is made from cellulose fibres created from wood pulp. To recycle it, metal and ink contaminants are removed and the process is simply reversed by soaking and agitating it, which releases the fibres that are then pulped. Cardboard can go through the recycling process four or five times before the fibres disintegrate fully. In its shredded form, cardboard can be used for animal bedding, stationery, and loft and cavity wall insulation. It has also found a final green resting place in the form of biodegradable coffins.

• **Food** disposal at source could be greatly improved. Only 34% of households in England compost their uncooked kitchen (i.e. vegetable peelings) and garden waste. Many authorities now offer a doorstep collection of cooked food waste for composting. A mix of food, kitchen and organic garden waste is put into huge anaerobic digesters where bacteria and microbes that thrive in such conditions break down the waste. Meat and fish should not be composted in home bins, as temperatures are not high enough to kill bugs that could become a potential health hazard, and scraps may attract vermin and pests.

• **Garden waste** accounts for a fifth of household waste in England. We produce the equivalent of about two wheelbarrow loads full per person per year. Fifty-five per cent of households in England have a doorstep collection service for garden waste. Generally it is sent to a processing plant where it's mixed with sewage sludge; here it breaks down, is composted, and used again as a soil improver. Woody cuttings, prunings, leaves and grass collected from gardens and parks are turned into mulches or outdoor surfacing material for paths and playgrounds.

• **Books** that have been placed in book banks are initially taken for sorting, where they are classified as either suitable or unsuitable for sale. The books suitable for sale can be sold in a charity shop. Books unsuitable for sale are generally recycled.

• **Household alkaline batteries** can now be recharged, increasing their product life. Most batteries, particularly lead-zinc batteries, can be completely recycled to make new batteries. The marketable products extracted from batteries during the recycling process include nickel, which is used in the production of stainless steel; cadmium, a component used in new batteries; and plastics, which are used in furniture production. Gold and copper are generally collected in very small amounts and re-used.

• **Car batteries** are broken up in a huge crushing machine and then sorted into their different parts. The plastic is recycled and made into various items including green recycling boxes, furniture, paint trays, car parts and drainpipes, and also turned into more cases to house car batteries. The lead is melted down for further car battery production, guttering, and shields for X-ray machines. The acid is treated and neutralised and the distilled water is purified and recycled.

• **Car oil** is decanted into a large holding tank where it is boiled and left to settle. Any water is removed and the oil is filtered. This process is repeated to finally produce a watery brown liquid used as an eco-fuel in power station furnaces and quarries. Reclaimed, treated oil is sometimes used as a mould release in foundries.

• **Cooking oil** is recycled into bio-diesel, which can be used to fuel specially adapted vehicles. It is also used as an industrial lubricant.

• **Textiles — re-usable clothes, shoes and household linens —** are collected in recycling banks and kerbside collections. They are often sorted by the organisation that collects them. Those suitable for re-use usually find their way into charity shops. Some are sold on to developing countries where they provide employment through refurbishment and retail activities, as well as affordable clothing and household goods for local people to buy. Other uses for textiles are as rugs and bedding in animal shelters It is important that all textiles be kept clean and dry otherwise they cannot be re-used. Ensure you tie any old shoes in pairs so that each pair stays together.

• **Paints** can be donated to the Community RePaint (CRP) scheme, a national initiative supported by ICI Dulux and funded from Biffa Waste via the landfill tax and the National Lottery. RePaint schemes organise collection of good-quality, leftover domestic paint from householders or trade decorators, and they re-distribute it free of charge to local charities, voluntary and community organisations, social service departments, and individuals referred by statutory and charitable organisations.

• **Electrical appliances** come in all shapes and sizes and are treated in various ways. Fridges, freezers and air-conditioning units have their cooling gases and insulation foam removed for treatment before the units themselves can be recycled. Other large units such as washing machines, tumble dryers, microwaves, and electric cookers are separated for metal recycling and recovery. Televisions, computer monitors and other equipment containing cathode ray tubes are sent for hazardous waste treatment before recycling any suitable materials. Fluorescent lamps and high-intensity discharge lamps (not household light bulbs/filament lamps) are also deemed hazardous waste because they contain small amounts of mercury. After the mercury has been removed for re-use or treatment, the metals and plastics are separated and recycled. Small electrical and electronic products, tools, computers, hairdryers, vacuum cleaners and so on are collected together for metals and plastic recycling.

• **Mobile phones** are often resold to raise funds for charities. They can also be refurbished and sold in other parts of the world, including developing countries. Mobile phones contain toxic elements such as mercury, cadmium, lead, and other substances that are landfill hazards. Re-use transforms a potentially toxic discard into a valuable commodity. Phones that cannot be re-used can be recycled for their metals and plastics.

• **Plastic bottles** are sorted mechanically. A machine sieves out any small particles and the bottles enter an auto-sort machine, which uses infrared technology to sort the mixed bottles into their separate types, e.g. high-density polyethylene (HDPE) or polyethylene terephthalate (PET), and also into their component colours. They are washed, then granulators chop the bottles into plastic flakes and density separation techniques purify the washed flakes. They are then ready for reprocessing. In theory, *all* plastic is recyclable, but only the plastic from bottles is suitable for this type of reprocessing. Food trays, yoghurt cartons and other tubs, although in bountiful amounts in most household waste, are not suitable and are rejected. HDPE flakes are reprocessed into drainage pipes on a large scale and used for motorway repairs, civil drainage and in agricultural markets. The flakes are also used to produce 'lumber' products such as decking and fencing. The PET flakes (polyester) are transported to manufacturers who use them to make insulation materials in cars or carpets. PET flakes are also used as the non-woven fibres for fleeces and jackets. Others are transformed into street furniture. Mixed plastics are recycled to make a wide range of products including flowerpots and watering cans.

25,000 plastic bottles weigh 1 tonne and recycling 1 tonne of plastic bottles saves 1.5 tonnes of carbon.

There are around fifty different types of plastics and they have literally hundreds of end uses; here are a few:

• **polyethylene terephthalate (PET)** — fizzy drinks bottles and oven-ready meal trays
• **high-density polyethylene (HDPE)** — bottles for milk and washing-up liquids and other detergents
• **polyvinyl chloride (PVC)** — food trays, clingfilm, bottles for juice, mineral water and shampoo
• **low-density polyethylene (LDPE)** — carrier bags and bin liners
• **polypropylene (PP)** — margarine tubs, microwaveable meal trays, bottles for a variety of products
• **polystyrene (PS)** — yoghurt pots, foam meat or fish trays, hamburger boxes, egg cartons, vending cups, plastic cutlery, protective packaging for electronic goods and toys.

Until recycling *all* plastics becomes commonplace, which I suspect is a long way off, we have to accept that disposal of some plastic items will continue to happen. Perhaps the most effective way forward is to try to eliminate it from entering our homes in the first place.

In 2006/07, 25.9 million tonnes of household waste were collected in England; 30.9% of this waste was collected for recycling or composting. The amount of household waste not re-used, recycled or composted was 17.9 million tonnes, a decrease of 4.2% from 2005/6. This equates to 352 kg per person of residual household waste and shows progress towards the 2010 target, in the Waste Strategy 2007, of reducing this amount to 15.8 million tonnes.

There has been a change in the composition of recycled waste over time. In 1997/8, paper and card was the largest component, making up 37% of the total, followed by compost (20%) and glass (18%). In 2006/07, compost was the largest component (36% of the total) with the next largest being paper and card (19%) followed by co-mingled (15%).

Research undertaken by WRAP to mark the start of Recycle Week in June 2008 highlighted five top items that people say they don't recycle:

• aerosol containers for deodorants, air fresheners, polish, etc (62%)
• biscuit and sweet tins (53%)
• plastic shampoo and toiletry bottles (32%)
• plastic bottles of cleaning products, e.g. bleach or disinfectants (31%)
• glass jars, such as those for cooking sauces, jams (20%)

The research also showed that 59% of people focused on recycling cans, plastic drinks bottles, glass bottles, paper and card. Over a quarter (26%) were unaware that items such as aerosols, plastic detergent bottles, magazines, and biscuit and sweet tins could often be recycled. In 2008 Recycle Week was about highlighting the extra things we can easily do to recycle more and to increase the impact of our recycling efforts.

According to Recycle Now, recycling one more thing could have a significant impact on increasing household recycling rates and help tackle climate change. If every household in England recycled one additional item from a range of common household products such as aluminium drinks cans, steel food cans, glass cooking sauce jars, plastic milk bottles and glossy magazines every week for a year, the total amount collected for recycling could increase by more than three-quarters of a million tonnes, and could potentially raise the national household recycling rate by up to 3%.

Recycling one more thing would also help tackle climate change by saving nearly 850,000 tonnes of carbon dioxide equivalent – the same as taking over a quarter of a million cars off the road each year, or saving the amount of energy required to power four million 32-inch plasma televisions for a year.

Rubbish obsolescence

After the Second World War, governments had to instil economic confidence and stability back into the countries that had taken a severe pounding. To boost reconstruction efforts we were encouraged to take out our hard-earned cash and go shopping!

The rationale was quite simple: spending money and oiling the wheels of commerce would bring necessary investment to manufacturing industries that had suffered great losses in wartime. People would get back to work, products would fly off the shelf, and the country would get back on its feet.

A startling modern echo of this was heard after the attacks on America on 11 September 2001, when President Bush reiterated the same message to a shocked and mourning nation. The President of the United States 'courageously and forthrightly called on the American people to go shopping' (source: www.iht.com/articles/2003/01/14/edspiers_ed3_.php).

This wouldn't be such a problem if we had a spare planet or three from which to pull in natural resources. In the past few decades we have used up a third of the planet's natural resources. Taking the example of paper and other wood-derived products, we are currently felling 2,000 trees a minute in the Amazonian rainforest alone. It doesn't take a mathematical genius to work out that, by continuing to feed our present hunger for consumption, we are soon going to run out of raw materials. If we don't act and adjust accordingly, we'll be running out of easy-to-manage solutions too.

The technical term for the making of stuff is 'the materials economy'. It's a very interactive system. It interacts with everything: ecological systems including animal and plant chains, the atmosphere, the oceans, you and me. Put simply, this interactive system has five key stages: extraction, production, distribution, consumption and disposal.

If you write those five words side by side in a straight line, you'll get some idea of the problem. It's a linear system. We live on a finite planet and you simply cannot run a linear system indefinitely or sustainably.

There's often little or no connection from disposal back to production, which means we have to keep starting again at extraction.

Before I explain how you're hoodwinked every day, let's bring another topic to the party: chemicals. Chemicals are used in just about every form of production, and some 100,000 synthetic concoctions are used in a multitude of manufacturing processes.

Did you realise that the humble cotton T-shirt has a startling tale to tell? Seven out of the fifteen most carcinogenic chemicals known to man are used to grow cotton. Do we consider this when we take one out of the packet and put it straight onto our precious skin?

Toxic chemicals are found in just about everything – garments, televisions, computers, mobile phones, batteries, shoes, shed paint and more. This becomes a serious issue when you look at turning that linear system I talked about into a more efficient and sustainable circle.

The toxic cocktail of a product's structure presents manufacturers with a great many hazards when it comes to breaking them down for re-use. It is a specialist process that is very carefully controlled but is in a relatively embryonic state. It is also utterly reliant upon consumers to feed it with product. We do this by depositing assorted spent items at the civic amenity sites and at the kerbside for the recycling rubbish collection truck.

Now, go back for a minute to the linear system referred to above and consider what's going to happen when the extraction process has no more raw materials. Does it sound like I've gone a bit sci-fi, sandal-wearing and irrelevant? I haven't, and unless you are in your late autumn years, you will probably be around to see the end of life with plentiful and cheap oil. It's a frightening notion, but perhaps still too 'out there' for us to digest in a heartbeat. I urge you to read *The Transition Handbook* by Rob Hopkins to shine a sobering yet optimistic light on the subject of a post-oil civilisation.

Maybe once we've actually lived through something so dramatic we will pay more attention, ditch the apathy, be more vigilant, and strike up a more positive attitude towards the no-brainer process we know as recycling.

Right, if I haven't convinced you with that fact, it's time to deploy another tactic. This is the bit where you sit back in your chair and say, 'You're kidding!'

Remember the idea of the linear materials economy? Imagine an enormous conglomerate that produced a product that was so good you never needed to replace it. If bits fell off or wore out, you could bolt a new bit on, and it would give you at least another good ten years of use.

Apply this idea to computers for example. It could never happen. Technology is out of date the nanosecond the first widget flies off the production line. Ask yourself why that is and you will find the answer is simple: the manufacturers have built in an element of planned obsolescence, or decay, into their products. They are probably capable of building a computer that would work perfectly well for many years.

They *could* make the components easily replaceable when software or hardware upgrades are needed, but they don't; they make you buy a whole new computer instead. They make keyboards that have a certain lifespan before the keys jam or the space bar doesn't work, and they make the new ones cheap, or sign you up to insurances that will pay for replacements.

They are selling you something knowing full well that the product has a limited shelf-life. You are buying into it every time you sign on the dotted line and sling your old model in the skip.

'Well, it just wasn't worth getting it repaired,' I hear you cry? Of course it wasn't! They designed it that way. They built in its keel-over date, manufactured it on the cheap, and sold it on the cheap. It is the same with many product lines — barbeques, trainers, phones, DVDs. They are all designed with the dump in mind. We're getting it at a snip, and Mother Nature is paying the dearest price.

It is estimated that up to three million domestic refrigeration units (fridges, fridge-freezers and freezers) are disposed of in the UK each year.

We live in a throwaway, take-away, replacement age, and we've forgotten how to mend and make do. We've been seduced by the companies who produce tatty products that temporarily satisfy our whims, and we don't give a second thought to closing the loop on a linear system we didn't even realise existed.

But that's not the end of it. There's another element to be aware of — perceived obsolescence. Perhaps better known as fashion, keeping up with the Joneses, call it what you like, the media machine perpetuates a constant bombardment of palpable pressure that beats down on us like a nagging in-law, telling us

to buy, buy, buy. 'Purchase these heels, honey. They're chic, they're really now, they're so you. Those flatties are so last year, sweetie, come on, keep up, what will everyone think of you wearing those/consuming those/using those/driving those?'

Designers dare not risk coming up with durable design, timeless beauty that will look great on you, not make your home look dated. Anyway, where would the fun be in that for Mr Manufacturer? It's much better to let you bring on the obsolescence yourself.

So, what's the solution to being sucked in by perceived obsolescence? Well, you can start by poking a finger in the eye of the system that's feeding us this propaganda in the first place. Break the chain by saying 'no' to consumption. Don't bow down to media pressure to own that product; your life probably won't be improved by co-existing with it and your pocket will be fuller by resisting it.

If you are going to make a purchase, why not buy something that has longevity built into it? Something that you will be able to get repaired when it goes wrong; something that is powered by a sustainable source (e.g. solar, wind-up), and not something that requires an expensive mineral-driven battery to run. In addition, consider taking the less toxic route. Organic isn't a fad; it's a solution.

Illnesses — from skin complaints to respiratory conditions and worse — are becoming more common, and the chemicals that surround us in everyday products and food are contributing to our collective health problems. Avoiding chemicals, where we can, is better for our health and the well being and welfare of those at the manufacturing end.

The materials economy and planned and perceived obsolescence seem embedded in our culture, but we can resist them and we can change the way we spend our money. Can you imagine what would happen if we all rebelled at the same time?

PART 2
House full of junk

Entrance ...
a welcome home ...

Room for improvement

Now you are moving into
a more positive frame
of mind about your
relationship with your
household waste, this is a
perfect room to begin with.
But before we get cracking
let's have a little chat.

The entrance to your
home can say a lot about
you. It's the first place other
people see as they cross the
threshold, and even those who
cast fleeting glances at it as
they whizz past every day (the
postman, the paperboy)
form an opinion of you
based upon it, like it
or not.

Everything you take into your home goes through
this checkpoint — all that excess packaging, the
wrapping, ribbons, string and bows, the polythene,
polystyrene, plastic, bubble-wrap that is allowed to
inhabit your personal space.

Somewhere between the arrival and the unpacking, sorting and
stacking of the contents comes the angst. What do you do with it all?
It can take as long to stuff the bag of bags with the spare bags that
you've accumulated as it did to unpack and deal with the contents.
And once you've liberated your fruit and vegetables from their
packaging, there's a pile of see-through plastic film, one of cardboard,
and another of that irritating scrunchy plastic that bounces back into
shape and invokes squatting rights in your bin. But remember: you did
bring it in.

Whizz forward to 3008. I wonder if the noble pursuit of
archaeology will have the same appeal as it does for today's intrepid
explorers, as they pace gently across digs, taking care not to disturb
the exciting finds waiting just under the surface of the dirt. I doubt it.
Let's face it, who's going to be interested in ferreting about in the crud
of Mr and Mrs 2008?

With your active consent, the entrance to your house is about to
undergo a dramatic change. It's time to start learning the many different
ways you can say 'No!'.

Unwanted mail is a horrific waste of paper. The waste paper thrown
into landfill every year could fill 103,448 double decker buses, and it
they were parked bumper to bumper they would stretch from London to
Milan. So let's make a start right here.

The first step can be taken in the time it takes you to boil the kettle
for a cup of tea. Grab a postcard or index card and write clearly,
'No flyers, circulars or leaflets required here — thanks!' and leave it
in view in a window by your front door, or stuck on the front of your
lockable mailbox.

Of course, this does require compliance from the person who's
trying to poke a message through your door. But on the whole this

method works well, and if it is abused you could always complain to the company the leaflet is trying to advertise.

You can also register not to receive any junk mail by contacting the Direct Marketing Association (UK) Ltd by letter, email or a quick telephone call and requesting that your address be removed from their system. You can also contact the Royal Mail Door-to-Door Opt-Out Service, and they will remove you from their lists for unaddressed mail.

Another way to reduce unwanted mail is to contact the Mailing Preference Service. The quickest way to register your details is online. Follow the simple steps to remove your details from mailing lists and to take off names of people who used to live at your address.

While you are there, you might also like to click on the button that registers your details to prevent unwanted telephone sales calls, which waste some of the most precious commodity of all — your time! If you don't have access to the Internet, contact the MPS by phone or letter (see Resources for details).

While you're waiting for your details to be removed from these lists, you can always write 'Unsolicited Mail — Return to Sender' on the envelope. The company who sent out the letter will have to pay a fee for the postage.

There is no one organisation responsible for the items that come through your door, which is why you have to go through all this rigmarole. However, if you do take these preventative measures, it should eventually make a difference. It takes up to six weeks to get everything in place, but soon the only things dropping through your letterbox should be things you want.

Another way to make a positive dent in your post is to go paperless on your household bills and statements. Banks, building societies, and many other companies welcome you using this service. However, while this is wonderful news for you, there will be a negative impact on your local post office, particularly if you no longer go in there to pay your bills. Be sure to use your local post office for other counter services and keep buying your stamps there too; they need your support more than ever before.

Change the habit, reduce the rubbish

• Use a local milkman and rinse and return your empties; this will save you having to deal with plastic bottles, which are trickier to recycle. Many milkmen also sell freshly squeezed juices, eggs, butter and other dairy produce. It might be a little more expensive, but if you don't live near the shops you should balance that against the convenience factor and the cost of running the car or taking public transport to nip out for the odd item of shopping.

• Speak to your suppliers and request that you go paper-free for your utility bills, banking and other regular statements.

• If you complete a survey, order or application form for anything, be sure to check the bottom for the statutory question of whether you would like to receive further information from them or their partners.

• As the mail drops through your letterbox, tear off the used stamps and put them in a handily placed recycled envelope kept near the front door. When you have enough, recycle them via Help the Aged charity shops, or send them to the RSPB (see Resources for details).

• When you get jiffy bags or any super-stiff packaging, open it very carefully and put it aside for re-use.

• When you receive your annual electoral register form, tick the 'Edited Register' option, which should keep your details out of the publicly available register.

Cash in on reduction

• LED and low-energy lights throughout the house will drastically reduce the amount of light bulbs or lamps that you have to replace.

• Keep a wind-up or shaker-powered torch by the door to light your path out from the door and for power cuts and fusebox emergencies.

• Buy a charity shop curtain to hang over the front door and ensure your letterbox and any other entrance points are well draught-proofed, which will cut out another terrible form of waste — heat loss.

• Don't discard shoes or trainers just because you've broken the laces. Keep a couple of pairs of spare laces in a drawer.

Reduce, re-use, recycle

• When you open your mail, do so with care. Re-use the envelopes and put a sticker over the address to give you clean space to write the new address. Write along the bottom, 'I am re-using this to help the environment,' to encourage others to do the same.

• Keep a recycling box by the front door for any unwanted paper, cards, junk mail or ripped-up envelopes. Put it out for collection on recycling day or drop it into a civic amenity site or paper collecting bin when you're passing.

• Use scrunched up newspapers and flyers as packing materials around delicate objects you want to pop in the post.

• If you or a neighbour has a rabbit or guinea pig, shred your unwanted paper and use it for animal bedding.

• Better still, opt out of receiving junk and unaddressed mail altogether (see above and Resources for full details).

• Keep a notice board by the front door and jot reminders on it of the things you need to do or buy when you go out next; this will help you minimise your shopping trips.

• Hang your cotton bags on hooks by the front door, so that you're ready to grab them when you go shopping.

• Invest in a pair of slippers and preserve your carpets from dirty shoes. It will also stop you wearing holes in your socks if you kick your shoes off as soon as you get through the door.

• Wash and dry used tin foil and place it behind your radiators to radiate the heat outwards. Don't place furniture directly in front of them.

• If you've collected a pile of useless spare keys throw them out with your recycled tins and cans.

• If you don't need the Yellow Pages, contact them online or by phone to be removed from their mailing list. You can find the most up-to-date listings on the Internet (see Resources).

• If you end up with more phone books than you need, you can rip the pages out, scrunch them up and throw them in your composter; alternatively, put them out for collection with your waste paper on recycling day.

• When sending delicate items to people, consider their rubbish bin too! Protect and pad out your package with popcorn. It's cheap to buy and make and completely biodegradable too.

Project box - Doorstep snacks

Hanging baskets and plant pots by the front door can be brought up to date and made delicious by filling them with enticing edible treats. They offer an eye-catching display and are a lovely nibble on your way in or out of the house.

Once you realise how simple it is to grow things like chillies and cherry tomatoes, you will be able to eat the most delicious fresh salad vegetables all through the summer. By growing your own, you also cut back on those wasteful plastic punnets.

For a colourful display, try planting cherry tomatoes with basil or coriander. How about baby cucumbers grown alongside a vibrant red chilli, or maybe bell peppers cultivated with dill or chives? Before you think I've gone mad, don't dismiss this idea. Try it for yourself. Chemical-free food tastes amazing, is better for you and your family, and you'll be so surprised how easy it is to grow.

Hanging-baskets or low planters are also excellent places to grow perpetual lettuce, of which there are many vibrantly coloured varieties with a diversity of tasty leaves. You can make a handful of lettuce plants last by picking off just what you need from the outside leaves, it will continue to grow throughout the summer.

Don't forget to harvest seeds from your crops and dry them out to plant next year. Share your abundance and give excess produce, plants or seeds to your friends and neighbours and encourage your street to be a foodie zone!

The only downside might be that your postman and paperboy spend a little too long admiring your front door ...

Encourage others to go paper-free. Here's an email to your local school, a good place to start encouraging others to have a go! If you have children they can send the email to their school and start a small paper-free campaign.

TO: Head@localschool.co.uk
FROM: Traceysmith@home.co.uk
SUBJECT: Opt-In Paper-Free Newsletters and Communications

Dear Headmaster/Headmistress

It is encouraging to see the school taking a strong interest in environmental issues, and Johnny keeps me regularly informed of your new initiatives to encourage and enthuse the children.

I'm sure you'll be pleased to know that our family are also reducing our impact on the environment by minimalising our personal and household rubbish output.

Following discussions with other parents, I would like to suggest that the school adopt an opt-in paper-free email scheme for the sending home of newsletters and general communications from school.

From the discussions I've had at the school gate, I am certain other parents and guardians would support the scheme.

It would be relatively quick and easy to put in place and, once the paper-free scheme was operational, it would save the school a great deal of money in paper, ink, printing, stapling, envelope and electricity costs, not to mention time saved by the school secretary in preparing and sending out such notes. In addition, you would forever eliminate the problem of children forgetting to take the letters out of their bags when they arrive home!

To get things rolling we would need to send out a note asking parents to join and requesting their email address. This can be typed into a simple database, or stored as a group of email addresses, ensuring none of the addresses are visible to others.

Then simply send out a test message, to which the parents should send a returned acknowledgement, and the system is up, running, and saving us all time, money and rubbish!

I look forward to hearing from you.

An eco-friendly parent

Utility ... clean up your act ...

Room for improvement

Our nation's obsession with housework and keeping things clean is insatiable. Advertisers shout at us from our television screens about the bluest shades of white and it seems there's a different bottle of jollop for every known nook and cranny. Putting aside the issue of the mass of bottles we have to find homes for, research tells us that we live too sterile an existence and that our children would benefit from exposure to a little more everyday grime.

The cornucopia of containers full of chemical products remains one of the biggest contributors to our weekly rubbish sacks. If we hark back to some of the old-fashioned methods adopted for keeping the house clean, we find a distinct steer towards Mother Nature's garden and the kitchen to source the materials. But you don't have to get out your pestle and mortar if you are not that way inclined as there are many multi-use environmentally friendly products available on the market, and there is now an encouraging shift towards having them refilled rather than buying a new bottle.

However, taking the non-toxic route to cleaning by making a few compounds up yourself isn't difficult. It will definitely save you money and storage space, will slim your bin bag and might even bring you softer and better-conditioned hands.

There are four key ingredients, probably all lying around in your kitchen as we speak, that could take care of many cleaning duties throughout your house. They are salt, white vinegar, baking soda and lemons.

You can bulk-buy some of the natural cleaning products — such as borax, soda crystals, bicarbonate of soda — that you can use for many different jobs. You could even go for a five-litre container of washing-up liquid like the one available from Bio-d (see Resources) and simply top up your normal-sized bottles yourself as required.

In my continual quest to reduce the rubbish my family and I generate, I'm always delighted when I can cut my travelling costs to get the shopping and increase the size of the vessel I cart home. I immediately reduce the rubbish I would have created by buying lots of smaller units. Some small local retailers will often give you an extra discount when you buy large boxes of things, in the hope that you will return to shop again.

There's a revolution taking place when it comes to laundry. You can now say 'no' to those bulky cardboard boxes or plastic bottles of washing powder and fabric conditioner by switching to soapnuts or eco-balls. They are chemical-free, great for washing, and designed to work well on short cycles.

Soapnuts seem bizarre when you see them for the first time and you really wonder how, and if, they are going to do the job. In India and Nepal, the soapnut from the plant *Sapindus trifoliatus* has been used as a washing detergent for generations. The ripe nuts are harvested in October and, when removed from the tree, are sticky and golden in colour. They contain saponine, a natural detergent. Place four or five half shells in one of the little cotton bags that come with them, tie the bag and place it with your laundry in the washing machine. Each tied pouch will usually give you two good washes and then you throw the spent shells into your composter to biodegrade. A 1-kg bag will cost about £12 and will clean approximately 100 loads of washing.

The eco-balls are quite different; they come in a box of three and look like little green plastic replicas of the planet Saturn, as they have a rubber ring around the centre. They are perforated and inside there are lots of free-flowing tiny balls that aerate the water and somehow 'bubble the dirt' out of your clothes. A set of eco-balls will cost around £25 and will give you around 1,000 washes; they even come with refill packs in the box. Soapnuts and eco-balls do not contain bleach or optical brighteners for artificial whiteness and neither will they strip your clothes of all their fibre, so you need little to soften them afterwards. A few drops of your favourite essential oil in the conditioner drawer of your machine will do a great job.

Adopting either of these methods will save you many cubic metres of cupboard space, and you will see an enormous reduction in rubbish carted away to landfill. You will also notice a dent in your shopping hours and energy spent in having to transport conventional soaps home (see Resources for details of both).

There are a growing number of places that sell your newly discovered eco-goodies, and your local health-food stores and organic product shops offer a perfect starting point. There are also a great many green stores competing online for your business, so do shop around for the best and most convenient deal for you (see Resources).

However, if you choose to shop locally, remember that by spending your money in your community, you are also putting much needed lifeblood back into the area that serves you in many more ways than the obvious. By being selective with where you spend your money, you can effect positive change, even if on a small scale.

It is estimated that 4% of the world's annual oil production is used for plastic.

We live in a society where we often feel dominated and intimidated by the supermarket giants. They are open twenty-four hours a day almost 365 days a year. We have come to accept them as a part of the fabric of our lives. They offer employment and claim to give us bargain prices, but with slurs of price fixings and mutters of monopolies, their darker side is coming to light as they continue to squeeze small retailers into submission and bleed some of our high streets dry. If you haven't tried it, you are yet to find out that shopping locally with a thoughtful and considerate retailer can be cheaper and makes you feel good too.

Change the habit, reduce the rubbish

• Don't use an extra scoop of washing powder for a pre-wash, simply pre-soak really grubby clothes overnight in a bucket with a scoop of borax or soda crystals, then wash as normal.

• Wash your clothes with soapnuts or eco-balls.

• Use essential oils to condition the laundry and for many other cleaning uses around the house. They bear different qualities; some are anti-bacterial, some anti-fungal, and lavender, lemon, spearmint, eucalyptus, sandalwood, pine and citronella are multi-purpose and some of the best ones to get started with (see **www.bookofrubbishideas.co.uk**).

• Salt, borax and bicarbonate of soda are inexpensive, easy to source, and are the basis for many key cleaning recipes.

Cash in on reduction

• Buy cleaning products in bulk wherever possible and, if you have too much, split the goodies with a friend or neighbour and share the shopping for the next batch.

• Use multi-purpose cleaners, which will deal with most jobs including sticky fingerprints on walls, stains on school blazers, silverware, copper, brass, chrome, stainless steel, pewter, leather, aluminium, acrylics, vitro-ceramics, tiles and grouting, to name but a few. Look for cleaners with no added phosphates, that decay biologically, do not harm the skin, and are safe for children.

• Buy magno balls for your dishwasher and washing machine. You won't have to buy packets or bottles of anti-calc treatments, as the magnetism of the balls changes the composition of the water and removes the offending elements that cause build-up and result in problems. Your appliances will also have a longer life span.

• Buy a bag-free vacuum cleaner and tip the contents into your composter to biodegrade.

• Use your utility space wisely and consider the tall, stacking storage units available to put your recycling in. See Resources for details of these and the other products mentioned above.

• Buy the handy guidebook, *Natural Stain Remover*, which is packed full of simple cleaning solutions.
• Keep your appliances, such as washing machines and dishwashers, serviced once a year by a professional; it's money well spent on prevention rather than cure.

Reduce, re-use, recycle

• Read the manuals for your major appliances and carry out your own basic interim maintenance on them every few months. De-scaling fluid isn't expensive and you could make your own. This would extend the life of your appliances and keep them out of landfill.
• Clear out all those surplus plastic carrier bags from your cupboards and use them as packing materials instead of bubble wrap.
• Collect together and send all printed and unprinted polythene wrappers and bags etc, from whatever source (provided all paper labels, sticky tape, residue, food remnants and foreign objects have been removed), to PolyPrint for recycling (see Resources).
• If you eliminate even a small proportion of the products currently sprawling out of your cupboards in this room, you will make a palpable dent in your weekly rubbish bag. But remember, if you throw away any bottles, jars and tins of cleaning products please do so responsibly. Do not tip them down the drain or empty their contents in the garden

(or anyone else's garden for that matter). Collect them all together, and next time you pass the recycling centre take them in and let 'the guys' do it for you.
• Pin up the leaflet from your local civic amenity site, with opening times and a list of the items they take, or visit **www.recyclenow.com** for a comprehensive look at what can be recycled in your surrounding area.

Project box - Eco-cleaning tips

Here are my top natural cleaning tips to help you get started.

Wooden chopping boards. Thoroughly rinse off any residue and rub rock salt into the grain with the cut side of half a lemon to sterilise the board and leave it smelling fresh.

Stainless steel. Give your sinks and other steel surfaces the shiny treatment by rubbing them over with baking soda and a damp cloth.

Mirrors and windows. Wipe them over with a 50:50 mix of white vinegar and water.

Chewing gum. Place the soiled fabric in the freezer for a few hours, then prise off the gum using the blunt edge of a knife. Dab off any remaining particles with eucalyptus oil on a cloth, then wash as normal.

Old wooden furniture. Wipe over with linseed oil, and allow it to soak in and dry before rubbing in a little natural beeswax to add a protective and decorative coating.

Kettle de-scaling. Fill the kettle with about half a litre of white vinegar and leave it to soak for an hour or two — do not boil this liquid! Then rinse and wipe out any residue, and rinse thoroughly twice more before making a cup of tea.

Lingering smells. In the fridge these are easily eliminated by inserting an egg-cup-full of baking soda. It's a great deodoriser and works quickly to absorb unwanted smells.

Gold. Clean any dull items with a non-toxic, sticky paste made from flour, vinegar and salt. Rinse well and buff up with a soft cloth.

Grubby grout. A toothbrush dipped in baking soda will improve the appearance of dirty grout.

Encourage others to recycle. Here's an email to send to your neighbours.

```
TO: Neighbour@nextdoor.co.uk
FROM: Traceysmith@home.co.uk
SUBJECT: Plastic bottle recycling proposal
```

Dear Neighbour,

I've just read *The Book of Rubbish Ideas* and have been thoroughly inspired to reduce rubbish in our street. As you know, we still don't have a kerbside collection for plastic bottles and as far as I understand, there are no immediate plans for one.

We've all been guilty of putting a few plastic bottles out for the 'normal' rubbish collection truck but I've just learned that they could take several thousand (yes, thousand) years to break down, if they are thrown into landfill!

I've decided I'm not going to let another bottle head to landfill from this house and I think I might have hit on a more efficient way of organising responsible disposal.

If a few neighbours all put our heads together, we could organise a weekly rota, so only one driver recycles the plastic from all six houses! It's the responsibility of the other households to deliver their bag of cleaned and squashed plastic bottles to the 'allotted driver' on Sunday morning.

That way, we only have one car on the road each week and we only have to do the run once every six weeks! That's petrol money and time saved and zero plastic bottle waste for our bin man!

Please find the proposed rota for the coming months below – I've already penned my trip in. I do hope you are up for the idea too and I look forward to hearing from you and seeing the rota fill up soon!

Deliver your bag of clean and squashed bottles to the allotted driver on Sunday morning between 9 and 11 am!

Tracey No 22	Duncan No 23	Izzy No 24	Gyles No 25
7/9 Tracey	14/9	21/9	28/9
5/10 Tracey	12/10	19/10	26/10

PS: If you'd like to borrow the book, pop round with your dates filled in and I'll lend it to you!

Kitchen ... cooking up a storm ...

Room for improvement

Recent government research has shown that public awareness of recycling has grown, with over half the population considering themselves committed recyclers. This is certainly a step in the right direction, until you touch on the subject of food waste.

Approximately one-third of the food we buy ends up in the bin. It's not just a question of how much money we've frittered away by letting this happen. More important is the considerable amount of greenhouse gases it omits once it lies festering in a rat-infested landfill.

New research published by WRAP shows that the cost of needlessly wasted food to UK households is £10 billion a year. The average household throws out £420 of good food a year, while the average family with children throws out £610. The study revealed that £1 billion worth of wasted food is still in date.
It costs local authorities £1 billion a year to dispose of food waste.

Current estimates say we waste around 6.7 million tonnes of food every year. It's hard to put that into context, as we generally have no idea what a tonne of food actually looks like. The following list might help.

According to the fact-packed Love Food Hate Waste campaign, every day in this country we waste:

1 million slices of ham

1.3 million yoghurts and yoghurt drinks

7 million slices of bread

5.1 million potatoes

1.6 million bananas

4.4 million apples

2.8 million tomatoes

Staggering, isn't it?

My tips list below concentrates on helping you find suitable solutions to this and other food-related problems. I hope it sparks your interest enough for you to buy a pinny and have some fun in the kitchen.

It's not just the food waste that is a problem. The packaging that surrounds, protects and brings home our shopping seems impossible to eliminate, but with some forethought and by adopting different shopping habits, you can cut it down to a pleasantly surprising minimum quite easily.

One of the biggest irritants for me is being told that we consumers are demanding this packaging. Nobody asked me! There are far too many assumptions being made, and they usually favour the supermarkets, not the customers.

If you feel aggrieved by this unwanted, un-recyclable plastic and foil, paper and cardboard, then make your voice heard. I've enjoyed writing an email and you can find it at the end of this chapter. All you have to do is copy it out and adapt it accordingly, then send it to your local supermarket prompting them to provide their customers with suitable receptacles to deposit unwanted waste at the point of sale. You may not know that, legally, customers are not obliged to take the packaging home, so why not leave it for the seller to dispose of responsibly?

Can you imagine the fuss it would cause if we started rebelling against the system and uniting the voices of thousands of consumers all over the country? One voice is a bit of a whisper, but a bunch of them can ring out like a choir! So get writing and do keep me informed of your progress and successes by clicking onto my blog, **www. bookofrubbishideas.co.uk.**

Supermarkets are everywhere. They are there for good whether we like them or not, and while I accept that they provide jobs and offer us cheap, convenient food at obscure times of the day, they have also done a great job of bleeding many of our high streets dry and putting small retailers out of business. I think that pushing a trolley up and down the aisles has to be one of the most soulless experiences known to man. However, even I use them from time to time. I am not perfect, I am human, but I'm also very selective about where I shop. I do not use supermarkets as much as I used to, and I feel a lot better for it.

Finally we come to those awful plastic bags. Over thirteen billion plastic bags are given away in the UK every year; let me write that out for you in full — 13,000,000,000. They are said to take between 400 and 1,000 years to break down, so the bags you binned last week could still be around by the time your children have had children and they've had children, followed by their children and their children, then their children, and by the time we get to their children the bags might be a quarter of the way towards complete degradation.

Plastic bags are harmful to the environment in many ways. Some blow free to be caught in the branches of trees and waft about like witches' knickers. Others make it to our waterways, killing fish, birds, mammals and many other land and underwater creatures who play vital roles in the food chain we depend upon.

The harm caused by plastic bags is so horrendous that I find it hard to believe that we haven't yet found a solution to this worldwide crisis. Our government has not taken the issue in hand or brought in legislation that will put an end to the problem. However, there is a solution. Just say, 'no thank you', produce your re-usable cotton bag and take your groceries home in that.

Change the habit, reduce the rubbish

• Get creative in your kitchen and start cooking with leftovers. Learn the fundamentals of soup-making and you'll never throw anything away again. Don't forget, you can always freeze it if you don't want to eat it that day.

• Portion control your dried foods like rice (portion for one child = 2½ tablespoons, four adult portions = one full mug) and pasta and, if you have an excess of these two starchy fellows, plan a similar meal for tomorrow, or use the leftovers in the next batch of bread you bake and reduce the amount of strong bread flour in the recipe.

• If your vegetables are looking as if they are on the turn, clean them, put them in water and pop them in the fridge to buy a few days' grace, or chop them, blanch them and freeze them.

• Use a composter, wormery or Bokashi bin to convert your peelings, teabags, eggshells, coffee grinds and other organic waste into a usable resource (see the 'Garden' section).

• Keep excess fruit and vegetables in a cold store or shed in a shady part of the garden until you need them. They will last much longer.

• Don't store bananas with any other fruit. The gas bananas omit encourages the other fruit to ripen more quickly.

• Buy dry pulses such as kidney, haricot and butter-beans and lentils. You need a far smaller dry portion size, and once you've soaked them overnight you'll find they almost double in size.

• Make stock from the carcass of your Sunday roast or the bones and heads of any delicious fish, and freeze it if you don't need it immediately. Stock is the perfect base for a gravy, sauce or soup.

• Put paper cake wrappers in your composter or wormery or, if they are exceptionally clean, throw them out with your other waste paper for the recycling collection. Alternatively, buy cake and muffin tins that don't require you to add a paper case, or use rice paper cases and eat the lot.

• If you are organising a children's party, start your planning by considering the rubbish it will create. If you don't want a stack of washing-up to do when it's all over, buy paper and biodegradable cups and plates and compost at home. Make finger food, or buy 100% biodegradable wooden knives, forks and spoons (see Resources).

• Stop buying supermarket pizza. It's much cheaper and far more delicious to make your own, and you have no polystyrene base, plastic cover and cardboard box to dispose of afterwards (see my step-by-step guide on **www.bookofrubbishideas.co.uk**).

• Be bold and join a growing band of people who remove wasteful wrapping from their shopping at the point of sale and leave it for the retailer to dispose of.

• Buy larger containers of things such as margarine and ice-cream and re-use the containers to freeze cooked items or leftovers to eat another day.

• Wash out and keep pretty glass jars for storing your own jams and preserves, or even buttons, bows and sequins for sewing projects.
• If you buy loose fruit and vegetables in a supermarket, stick the label straight on the items if appropriate. Avoid sealed packs and punnets of produce wherever possible.

Cash in on reduction

• Join a local vegetable box scheme, preferably organic. The most scrumptious, seasonal food will be brought to your door, usually with a recipe sheet to help you cook more unusual items, and it won't be wrapped in copious amounts of plastic. The empty box is collected when the next full box is delivered, and you will save time and petrol, or other travelling costs, by not having to shop for it. See Resources for details of vegetable box schemes, or visit **www.bigbarn.co.uk** and enter your postcode for a list of schemes and small local producers near you.
• Get a couple of FreshPods. These are organic compounds that come in small plastic cases and sit in your salad tray in the fridge. They absorb ethylene, the gas that food produces when it biodegrades, and therefore your fresh produce lasts longer.
• If you can afford to, buy a biomass boiler that will burn wood and all other burnable household waste to heat your house and your water.
• Drink tap water or, if you absolutely must, buy a water filter (see Resources). But stop buying the plastic bottled stuff.

Reduce, re-use, recycle

• Store lemons in a separate fruit bowl to use for cleaning and cooking.
• Give old pairs of spectacles to your optician for recycling. Most reputable opticians have a recycling scheme, or visit **www.biggreenswitch.co.uk** for details of schemes near you.
• Send dead batteries to Energizer for safe recycling (see Resources).
• Be kind to the people that sort through your recycling and take a moment to rinse out all your cans and bottles.

• Cut up old plastic containers into strips and use an indelible pen to mark them up as plant labels.
• There are several recycling schemes available for disposal of old mobile phones and printer cartridges. Reclaim-it (see Resources) offers a complete fundraising system and ongoing support to help you develop your own fundraising scheme, focusing on collecting empty printer cartridges from friends, colleagues and businesses in your local area.
• Put cut hair and pet hair into your composter.
• Maximise the space in your recycling bin by squashing your cans, tins and bottles by foot or by using a multi-purpose can crusher. By doing this you will need to make fewer trips to your recycling centre.
• Keep an old tin can to use as a cookie, pastry or scone cutter, instead of plastic ones that eventually get brittle and break.
• Re-use an old tin can or a cardboard coffee or hot chocolate container as a pen or pencil holder.
• Do you have a collection of obscure foreign currency in a drawer? It's time to have a clear out and drop your dosh into a local charity shop. Many popular high-street names take donations, including Danardo's, Help the Aged, Marie Curie Cancer Care, Sue Ryder; they'll be grateful for every dime.
• If you cook duck or goose, sieve the liquid fat to remove any bits, then store it in a jar in the fridge where it will solidify. Use it on your next batch of roast potatoes; they will taste delicious!
• Empty your hot-water bottle into a watering can to use on your plants.
• Make upside-down planters from large plastic milk cartons to grow tomatoes. Watch my film on **www.bookofrubbishideas.co.uk** to see how to make these ingenious devices and teach your children a lesson in cultivation, preservation and recycling all in one go!
• Boil up any vegetable scraps and potato peelings for your chickens to enjoy and convert into wonderful free-range eggs.
• Don't use cling film or kitchen foil to wrap sandwiches. Invest in re-sealable boxes to store lunches, homemade yoghurt and other yummies.
• Instead of buying packet refills, grow, dry, chop and store your own herbs, or freeze blocks of them into ice cubes to keep fresh.

Project box - Bird bites

If you've cooked a fatty joint of meat, or have been frying in dripping, butter or lard, here's a simple way to get rid of the fat and give your feathered friends a treat at the same time.

If you only have a small amount to dispose of, let your frying pan cool down, take a slice of bread, mop it up and put it straight out for the birds.

It you have a large amount of fat to deal with that will eventually set hard, why not make up a yummy fat ball for the birds to enjoy? Pour the liquid grease into a bowl and add mixed birdseed, peanuts, any chunks of leftover cake or stale breadcrumbs, a few raisins and stir well. Tie a piece of string around a small twig and place it, twig first, into the bottom of a suitable receptacle such as an old yoghurt pot, then pour in the mix and press it down, keeping the string in the centre of the pot. Leave the mixture to cool down completely and place it in the fridge overnight to set.

Remove the birdie snack from the pot and tie it somewhere where you can observe and enjoy from a comfortable viewpoint.

Yoghurt - Pots of goodness

• Heat 500ml of full-fat or semi-skimmed milk to boiling point, then take away from the heat.
• Stir in 1tbsp of natural yoghurt.
• When cooled a little (about ten minutes) pour into a large wide-brimmed flask, put on the lid and leave for about fifteen hours.
• Then add whatever fresh or dried fruit you have to hand; banana chips work well and stay fairly firm. It sometimes is a little lumpy. If this frightens you, just stir it and the lumps disappear.

For added smugness, use the empty carton from the natural yoghurt you had to buy and plant a seed in it! Grow your own strawberries to go with your yoghurt! Oh, and when you make the next batch, just use a tbsp of the yoghurt you just made. There is no need to buy natural yoghurt ever again!

Here's that email for your supermarket ...

TO: localsupermarket@town.co.uk
FROM: Traceysmith@home.co.uk
SUBJECT: Your Corporate Social Responsibility obligations and disposal of unwanted excess wrapping and packaging.

Dear Mr/Mrs Supermarket Manager

Recently, I learned that packaging waste makes up around 24-30% of the household rubbish generated throughout the UK. I also discovered that packaging represents £10 (13%) of each £75 spent on the average shopping bill.

The responsibility for disposal of this packaging is immediately passed to me the moment I purchase goods and leave your store. However, I am not obliged to take this excessive packaging home with me, and I feel the time has come for change. If the packaging stays within your shop, the onus falls upon you to dispose of it responsibly.

I would like to see suitable receptacles in your store for customers to deposit their excess packaging in. We will require somewhere for cardboard/plastic film/cellophane wrapping/multi-pack holders etc.

Worldwide environmental issues are a concern for us all and the profile of recycling is being constantly elevated.

I fail to see why consumers should continue to have to deal with a problem that has been created by retailers. It's time for you to accept your corporate social responsibility and show us that you care about recycling.

I have been a loyal customer of your store for many years and I look forward to hearing from you as to when you will be installing these facilities.

The receptacles need not be custom-built - simple large boxes will suffice in the short term - but please look seriously at taking this initiative on in a permanent form. If we as a nation are going to make any positive impact to our local and global environment we must act now.

If I do not receive a reply to this email, I have no choice but to leave my excess wrapping and packaging at your till, as I am perfectly entitled to do. I have many supporters of this simple initiative who are also willing to act.

I await your reply with hopeful anticipation.

cc: Your local radio newsdesk

Study ...
turn over a new leaf ...

Room for improvement

The study is an interesting and eclectic pocket of space. It is often crammed with more technology than a suburban branch of Curry's, and you usually find a melange of computers, speakers, televisions, musical equipment, clocks, printers and various other gadgets, as well as unidentified cables and leads that charge up something.

The options to dispose of working but unwanted equipment of this sort often leads you to the doors of eBay and other auction sites, local car boot sales, the free ads in your local paper, 'for sale' cards in the sweet shop or supermarket, Freecycle and, if it's not working (and more depressingly when it still is), the local civic amenity site.

Electronic waste, or E-waste for short, is the stuff that landfill nightmares are made of. A European Union directive in August 2005 called the WEEE (Waste Electrical and Electronic Equipment) directive required all producers to take responsibility for their discarded products. This means that if you purchase something of an electrical nature, at the end of its life the retailers now have an obligation to take it off your hands and to organise the safe and responsible recycling of it.

This directive concerns not just office equipment, but larger household items such as fridges, freezers, washing machines, electric cookers and almost everything else with a plug on. This is one of the fastest growing waste streams in Europe, and in the UK we are responsible for dumping approximately 1,000,000 tonnes of the stuff every year.

Not only does the quantity of metal and plastic from these items take up valuable space in landfill sites, but toxic substances and harmful gases are omitted during its degradation. (Let's face it, these things are never actually going to biodegrade.) This has serious implications for the state of our health and the volatility of our climate for centuries ahead.

Before I explore the options open to us for disposal, this seems like an appropriate time to ask you whether you actually need to buy a new gadget in the first place. Would a software upgrade be possible? Is there still life in the product and could somebody else benefit from using it? Could you donate it to a charitable organisation to help those less fortunate?

I've already covered the materials economy at the start of this book and I've done my best to expose the fact that we are constantly enticed to buy, buy, buy, but the consequences of this are far reaching. Perhaps this really is an appropriate juncture for you to start saying, 'No, I don't really need a new widget, this one does a perfectly good job.'

If I can't tempt you to be that radical, let's look at what options we have to comply with the WEEE directive. There is currently no unified system in place and I imagine this will remain the case due to the sheer logistics of it all.

We can do what many responsible householders have been doing for some time and continue to take dead equipment to the civic amenities sites. Or, if we make an electrical purchase, we can ask the retailer how he is going to collect and dispose of our old item. They are obliged to do this free of charge if you make a like-for-like purchase, regardless of where you purchased the old item. This is easy to do for small things such as radios or little televisions, as you can probably arrange to drop them off at the shop when you go in to buy your replacement, but for big items you may have to arrange a collection, which may incur a charge.

The same goes for your local council. They are obliged to remove your old goods, but you may be charged for the privilege. For further information, speak to your electrical retailer, your local council or get in touch with the Environment Agency Helpline (see Resources).

Enough of this seriousness; the study is also a place where nice things happen! I rarely remember to send birthday cards. I usually send out a belated wish well after the event. However, when I do get my head in order, I love to make my own cards.

Shop-bought cards are amazingly expensive. Do this little exercise and you'll be in for quite a shock. Tot up the number of your immediate family and times it by two. This takes into account Christmas or another annual celebration and their birthday. Multiply that by £2.50, the average cost of a card. Bear in mind that you have not allowed for any neighbours having babies, friends passing their driving test, chums at work retiring or, if you have children, any of their friends celebrating anything! It's a staggering figure.

If you sit down with a few basic materials and some of your 'rubbish', you can make a dozen cards in one go. I guarantee that the recipient will remember your card above all others because you took the time to handcraft it. Look in the Project box for a few simple ideas and plan an afternoon of sticky fun sometime soon.

Change the habit, reduce the rubbish

• Put all of the interesting bits and pieces from your rubbish in a designated art box and use them for making cards and wrapping and tagging gifts (see the Project box). Check out more ideas on my film at **www.bookofrubbishideas.co.uk**.

• Only use your printer when absolutely necessary.

• Print on the back of waste paper if you can.

• Print in draft where possible.

• Print double-sided where possible.

• Use refillable toner/printer cartridges and drop your old ones off at the recycling centre, or check your high-street retailers; many have special receptacles in store.

• Go green and email your greetings cards! It's not mean, it's savvy. There are many free sites, and you can create interactive cards. One of my favourites is **www.jigzone.com** where you can upload a photograph of yourself, turn it into a jigsaw and send it as an e-card.

• Wrap your presents in simple brown paper, tied up with string. You can buy a big roll of brown paper; it's cheap and you won't end up with half-used sheets of expensive wrapping paper in cellophane wrappers, which eventually find their way into your bin.

Cash in on reduction

• Buy sticky labels (there are many worthwhile causes to support in this, e.g. Friends of the Earth or Baby Milk Action; see Resources) to put on old envelopes so that you can re-use them.

• Buy a stapleless stapler and you'll never have to buy boxed metal staples. They are more effective than you might expect.

• Buy charity gifts — a goat or a donation to a school project — as presents for your friends and family and you'll have nothing to wrap up at all.

• Go to **www.charitygifts.com**, a shopping portal that lists all the wonderful things you can buy that can make a real difference to our world.

• Buy a hand-operated shredding machine to deal with sensitive letters and bank statements and put the shreddings into your composter to balance out the wet waste.

Reduce, re-use, recycle

• Choose refillable pens. It is estimated that plastic biros will last in landfill for 50,000 years and daily worldwide sales of biros exceed some 11,000,000 units.

• To help close the loop of the materials economy, we must obviously recycle our goods, but we must also buy recycled goods.

• Subscribe to magazines that you can read online, such as *The Ecologist*.

• Use the library to source books that you can change regularly.

• If you are not going to use the CDs and DVDs that are given away with your daily papers and magazines, either drop them into a charity shop or use them as bird frighteners in the garden or on the allotment.

• Why not send your old mobile phone to **www.envirofone.com**, a charitable project that runs a phone-recycling scheme?

• If you have half-used packs of paper, envelopes, pens and other bits of stationery that are clogging up your desk, donate them to a local charity such as NACOA (the National Association for Children of Alcoholics).

Project box - Crafty with rubbish

All you need to experience your own 'Blue Peter' moment is listed below, with a few ideas on what you can use to decorate the cards.

You will need:
• A box to keep all your bits and pieces in.
• Plain cards and envelopes to match, ideally a minimum card weight of 270 g for the cards.
• A glue stick, a ruler, scalpel or scissors.
• A plastic wipe-down tablemat to protect the table.
• Patterns cut out from a magazine or catalogue.
• Old gift-wrapping and tissue paper.
• Buttons and toggles.
• Sweet wrappers, used, clean tin foil and netting from fruit bags.
• Beads from broken necklaces and bracelets.
• Leaves or pressed flowers.
• Bows from presents you've received.
• Bits of wool, string, raffia, ribbon and rubber bands.
• Crisp packets, corks, wooden lolly sticks.
• Stale pulses such as lentils, coffee beans and spaghetti.

Looking for inspiration?
• Cut a small piece of old towelling, shape it into a nappy, and fasten on the front with a big safety pin.
• Cut up old photographs or postcards and make a collage or write speech bubbles above them.
• Corrugated cardboard from inside perfume boxes makes an interesting and smelly statement on plain card.
• How about lip-sticking your lips and planting a kiss on a plain card?
• Cut out lots of wedding images from a magazine and stick them on randomly to make a wedding card.
• For gift tags, use any stiff card and attach it with string or ribbon; decorate with the same embellishments as your greetings card.
• Buttons and toggles from old clothes can be tied onto the ends of labels and tags.

While you are in your study, how about sending an email to your local authority about the waste collection and recycling in your area ... here's my attempt.

```
TO: localauthority@town.co.uk
FROM: Traceysmith@home.co.uk
SUBJECT: Keeping organic waste out of landfill
```

Attn: Household Waste and Recycling Departments

Dear Local Authority

I live in ..., alongside the neighbouring county/district of ... At present, we do not have a recycling collection for organic food waste and I am not aware of a scheme starting anytime soon; please tell me if there is one in the pipeline.

I am aware that DEFRA does not support a unified policy on recycling facilities. However, I understand they do strongly encourage neighbouring counties and districts to 'work together' for the common good and share facilities where possible.

In nearby ... there is a weekly/fortnightly collection of food waste that is taken to a nearby anaerobic digesting/composting facility and I am writing to ask that this service be extended to our county/district too. It may take some organising, but surely cannot be too difficult to implement as the processing plant is already in the vicinity and unlikely to be working to full capacity.

Many people in my local community would support this initiative and, no doubt, there are countless more in the surrounding area. I attach a petition to show the united support for this idea.

I have copied this email to our local BBC radio station and our MP ... who I am sure will follow this story with great interest, as they also seem very keen to encourage sustainable-living behaviour. I am certain that a shared scheme for responsible disposal of biodegradable organic waste would be seen as a very positive step in the right direction for our council and I look forward to hearing from you soon with your thoughts on putting this idea into action.

cc: Your local radio station
cc: Your local Member of Parliament

WC ...
flushed with success ...

Room for improvement

With the important issue of size in mind, and with this space
being so small, you might think there's no waste in a WC
apart from the obvious human variety. In the smallest
rooms up and down the country you are likely to find
toilet paper, maybe some reading material, soap,
a hand towel, probably a spray that you hope will
restore a neutral smell and something that kills many
known germs.

Toilet cleaners quite often contain hydrochloric acid, which is a
highly corrosive irritant to the skin and eyes and can cause damage
to the kidneys and liver. They can also contain hypochlorite bleach,
a corrosive irritant that can burn eyes, skin and cause damage to the
respiratory tract. Incidentally, it is estimated that sodium hypochlorite
solutions cause 3,300 accidents needing hospital treatment each year in
British homes (RoSPA, 2002).

Thankfully, an expanding range of very effective eco-cleaners is now
available. They use natural, plant-based, non-toxic ingredients, which are
less damaging to our waterways and the environment. You can take your
empty bottles along to many local health food shops and stockists and
they will be happy to sell you a refill.

Tins filled with chemical air-freshener sprays are quite simply
ridiculous. They fill the air with an ultra-fine mist that you inhale into
your precious lungs, usually in a very confined space. Regardless of what
the manufacturers say is in the spray, it really is not a sensible thing to
use, unless it's filled with pure fresh air — which it isn't.

Almost all of these sprays comprise chemical concoctions derived
from petrochemicals that have been linked to many health problems.

A quick look at the label on the back of the tin will give you some idea of what's in there, and I can guarantee you'll find a 'caution, danger or irritant' sign on the back too. Bear in mind that the manufacturers rarely list all the ingredients. Even if they did, would you be likely to read through them and research them on the Internet or in your local library? Possibly not, but now you've read this page I certainly hope I've stimulated you to do so.

A short delve into this complex subject will show you that some of the chemicals are alleged to be responsible for sore throats and runny noses, skin complaints, sinus problems, nausea, painful muscles, nasal congestion, wheezing, shortness of breath, not to mention being a common trigger for asthma, headaches, mental confusion, listlessness, depression, seizures, inability to concentrate, irritability, restlessness and conversely, sleepiness.

Bear in mind that airborne sprays can also be absorbed through our exposed pores and children have much finer layers of skin than we do. It's also much harder to get something out of the body that has permeated in this way. So, how to freshen the air? Open the window!

Change the habit, reduce the rubbish

• Replace aerosol spray cans with pot pourri; see my recipe in the Project box.
• Keep a lemon balm or other scented plant in the room.
• Bars of soap can harbour unseen germs and are often discarded when tatty even though they have plenty of use left in them. Instead, use a more hygienic hand-pump dispenser that you can refill.
• The Australians have a great saying, 'If it's yellow, let it mellow, if it's brown, flush it down!' You do not have to flush every time after taking a pee; it's a waste of water, which is a precious resource. If I cannot convert you, at least use a flush resistor such as the Interflush and reduce the amount of water you use. Flush resistors will also inhibit the amount of toilet cleaner that's dispensed from any clip-on devices and reduce the frequency with which you have to replace them.
• If you use the clip-on toilet cleaning blocks, invest in a refillable one.

Cash in on reduction

• Ideally, buy white 100% recycled toilet roll; white paper requires less processing during manufacture, is usually cheaper, has no coloured dyes, biodegrades easily and, if the demand continues to rise, other brands that use virgin materials will also have to fall in line and source recycled paper.

• Buy your toilet paper in bulk and, if your household doesn't use much paper, go 50:50 on a huge bag with a good friend or neighbour and take turns in buying it and other bulky cleaning products.

• Buy cleaning products in the largest available sizes and top up the smaller bottles kept in your WC.

• Pop an eco-toilet de-scaler into your cistern to prevent limescale build up (you may need two in very hard water areas). They are chemical free and your toilet bowl will be clean for up to five years.

• Do all these things and you'll hardly have to go shopping.

Reduce, re-use, recycle

• Clean the sink and taps with a damp cloth sprinkled with a little baking soda for a shiny sink and gleaming taps.

• Use the cardboard part of toilet rolls in your garden as biodegradable holders for seedlings or small plants.

• Or throw them, along with any other cardboard rolls, such as from kitchen paper, into your composter or wormery to balance out the 'wet' waste (see 'Garden', pages 96–97, for essential advice to virgin composters/wormers).

Recycling one tonne of newspaper will save seventeen trees, four barrels of oil and 4,200 kilowatt hours of energy.

Project box - Coming up roses

You can save money and reduce your overall bottle count by making a natural version of your toilet cleaner. Mix 1 cup of borax with ¼ cup of white vinegar or lemon juice and combine to make a paste, apply to the inside of the bowl and leave for a couple of hours. Scrub with a brush and you have a clean toilet at no risk to your health. Splash a few drops of cleansing tea-tree or lavender essential oil on a tissue and wipe around the porcelain rim of the throne as an anti-bacterial aid; the used tissue can be safely sent down the U-bend next time you flush.

If it's intolerable to be in your privy without a pleasant scent wafting gently about your person, consider using pot pourri sprinkled with a few drops of your favourite essential oils.

I recommend a nice, spicy mix to help you deodorize this room. To make your own, first gather handfuls of flowers from your garden or leaves and deposits from the trees from a nearby woodland walk. Include some twigs and pine cones too if you can; they are great for holding the scent of your favourite oils. Dry out your chosen mix out by placing it somewhere very warm, or on a shelf in direct sunlight for a few days. When it's dry, throw it in a large jar or bowl, intersperse layers of it with salt (rock salt looks quite funky) and add drops of the oil of your choice. For a spicy version, why not try cinnamon oil, or a couple of cinnamon sticks, orange oil, sprigs of juniper with berries, a couple of sprigs of rosemary (run your pinched fingers over the leaves to release the oil) and a couple of bay leaves. A few drops of peppermint oil works well.

This homemade pot pourri will absorb unwanted smells and you can refresh it with 'safe' new drops whenever you feel the need. No more cans, sprays, plug-ins, click-downs or strap-ins to fill up your dustbin with superfluous rubbish!

Living and Dining ...
food for thought ...

Room for improvement

In the UK around 4,000,000 children live in households
that cannot afford to replace worn out or broken furniture.
The Furniture Re-use Network (FRN) was established in 1989 to bring
together social, economic and environmental benefits through the re-use
of unwanted household items. The FRN presides over 400 recycling
organisations who collect and redistribute furniture and electrical items
throughout the UK to people in need. Over 2,000,000 assorted pieces
of furniture have been passed onto low-income families, and this keeps
around 90,000 tonnes of waste out of landfill each year. If you want
to donate or receive furniture, visit **www.frn.org.uk**, which will guide
you to a local agency. Objects you no longer need, and may consider
'rubbish', may be another's treasure.

If you are Internet-savvy, **www.freecycle.org.uk** can help you to
find a new home for items you no longer want. Their mission is 'To
build a worldwide gifting movement that reduces waste, saves precious
resources and eases the burden on our landfill sites, while enabling
our members to benefit from the strength of a larger community.' The
process is as simple as the concept. Visit the website to join your local
group, and list the stuff you want to give away. Interested parties will
send you an email to say, 'Oh, that would come in handy, how can I
arrange to pick it up?' and you take it from there.

Just about everything can be found on Freecycle. I have Freecycled
almost all of my baby furniture, an old Ford Escort car, a wooden tea
chest, a bread maker, light fittings, an old washing machine, a tumble
dryer that was on its last legs, and various other bits and pieces. I used
it when my mother passed away and I was left to clear out her little flat
over an Easter weekend when all the charity shops were closed. I gave

away every stick of furniture that wasn't bolted or screwed down. The items went to an assortment of people, including some who were broke and others who were in the process of splitting up with their partners and trying to set up a new home. Knowing that mum's stuff was going to a new home left me feeling warm inside.

I've also been on the receiving end of Freecycle. When my family and I spent Christmas with my friends they put up a 'Wanted' note for five chairs so that we could all sit down and eat dinner together! As soon as Christmas was over, the chairs were Freecycled to somebody else.

FRN, and Freecycle in particular, restore my faith in human nature, and the volunteers that run their groups are amazing (see page 117). If you cannot find one in your area, perhaps you could consider starting one.

Change the habit, reduce the rubbish

• Contact the Furniture Re-Use Network or join your local Freecycle group and give away unwanted furniture and household goods.
• Make your mealtimes more fulfilling and enjoyable and reduce your rubbish by ditching pre-packed ready meals in favour of simple, home-cooked dishes.
• If you've never experimented with cooking, borrow cookbooks from the library or buy some from the charity shops.
• Recycle, Freecycle or sell the unused electrical gadgetry that's taking up valuable space in your cupboards, such as that fondue maker, popcorn maker, sandwich toaster or carbonated-drinks maker. If you haven't used them in the past six months, you are unlikely to use them in the next six months. Do it before you end up slinging them all in the bin in a moment of frustration and madness.
• The dining room is a prime target area for anyone buying newlyweds a present. You don't want to end up with three toasters, so make your wishes clear about the gifts you'd like by using a gift-list service, or ask for money or vouchers and buy exactly what you need.
• If your fruit bowl is full of limp and unappealing apples and pears, don't throw them away! Chop the fruit up and make a quick fruit salad, stick it in a juicer, or stew it and make a purée to go over ice cream.

• If you have bread left over after dinner, don't throw it away. Learn how to make a delicious bread-and-butter pudding for tomorrow's dessert.

• Don't decant jams, marmalades and butter into dishes for the breakfast table. If they aren't completely used up, you often end up throwing the remainder into the bin, which is a complete waste of good food.

• If you like to see flowers in your rooms, ditch the bouquets and go for pots of everlasting flowering plants instead and swap them around the house when you fancy a change of colour and smell.

• Forget all those boring party-plan evenings; host a clothes swap, or kids' toys and clothes party instead, save yourself a fortune and have a great night in with your chums.

Cash in on reduction

• Book yourself onto a cooking course and learn how to make a few basics, such as bread or soup, which could have a considerable impact on the pre-packed grocery rubbish you amass.

• Don't buy salt and pepper cellars that you can't refill.

• Use washable cotton napkins and tablecloths or, if you must use paper ones, dispose of them in the composter or wormery.

• Find an old solid wooden dining table complete with knocks and scratches from a local auction room. It's likely to be cheaper than a brand new one and you can always bring it back to life with sandpaper and a protecting coat of natural varnish or oil.

• Instead of using tea-light candles to bring a moody glow to your romantic dining table, consider low-energy lighting with tabletop LED lamps such as the Candelas from Vessel (see Resources). These are more cost-effective than tea lights, and safer too. They give off a beautiful ambient light and you'll have no more metal casings or plastic holders to dispose of.

• Stop buying boxes of tissues and plastic packs of throwaway pocket

hankies. Invest in some cotton handkerchiefs and wash them instead.
• To warm your room, why not invest in a wood-burning stove? It will also make good use of your spare paper, newspapers and cardboard.
• Don't be drawn into the 'buy now and pay on your deathbed' culture. Consider getting your old sofa re-upholstered instead of throwing it out and replacing it. Some colourful fabric and a couple of reinforcements to the springs could give it a new lease of life and cost you less.

Reduce, re-use, recycle

• Keep your meals warm with inexpensive tea-light-operated warmers or, better still, get everyone to sit at the table at the same time to eat!
• Stop buying and serving bottled water. Tap water is perfectly safe and you can filter it if you want to.
• If you heat the house with a stove, make more use of the top of it to cook or warm through food and don't waste your precious gas or electricity. It's an ideal place to prove dough and the smell of fresh bread cooking is wonderful.
• Another great waste found in this room is your precious time! Involve your children or partner with the laying and clearing of the table. The cook is encouraged to do more cooking if he or she receives more help, and the whole job is easier if you are working as a team.
• If you've broken half of your crockery set and no longer use it, donate the remainder to a local charity shop and do somebody else a good turn.
• Put any broken crockery to good use by putting it in the base of plant pots in the garden; it improves drainage.
• Be sure to recycle all glass bottles from your drinks cabinet. Not enough clear glass is put into the recycling banks. We produce plenty of clear glass in the UK, but a lot of it is exported as bottles for spirits. By comparison, we import a large amount of green glass, principally as wine bottles, which we do manage to recycle. Green bottles made in the UK contain at least 85% recycled green glass.
• The dining room often doubles up as a playroom and bits of old jigsaw puzzles regularly end up under the sofa. Throw any incomplete boxes out with your paper waste.

• Stop by a nearby farm shop and support your local economy by buying loose fruit and vegetables grown in the area; put them in your cotton bag to transport home.

• Keep your entertainment collection fresh by using your local library to borrow computer games, CDs and DVDs instead of buying new ones. They usually have a huge selection, including many of the old classics. You can hire them out for a week or a fortnight at a time and they are often cheaper than the local video shop. Make enquiries at your library to see if you have a mobile bus that stops near you, instead of driving into town.

• Your old vinyl may be lurking in the living room. Open an account on eBay and sell it on; there are many international collectors of old singles and albums, and records from the 1960s are very popular.

• Your old Sky digiboxes can be recycled at no cost by sending them to: Freepost RLUT-GCLR-LBXK, Unipart Technology Logistics Unit G, Swift Park, Old Leicester Road, Rugby CU21 1DZ.

About one-fifth of the content of household dustbins consists of paper and card, of which nearly half is newspapers and magazines. This is equivalent to over 4 kg of waste paper and card per household in the UK each week.

Project box - Bric-a-brac

We've taken care of the big items of household furniture, but sometimes all the irritating little bits of useless rubbish can take up just as much space. However, *your* idea of rubbish might not be the same as somebody else's, and organising a car boot sale can be a great way of purging your entire home and making a few pounds in the process.

Car boot sales can be great fun and, if you are a bootie virgin, I recommend you do one with a good friend and clear out two houses in one go! Take a look at your local paper to seek out a good venue or, if you want to go further afield, visit **www.carbootjunction.co.uk** who have information on hundreds of sites around the UK. Find yourself a reinforced pasting table to borrow for the day, take a flask, some sandwiches and plenty of change for your prospective punters.

If you don't wish to raise funds for yourself at a car boot, the nick-nacks you have may raise funds for others. Fundraising events are held throughout the year, and schools are constantly crying out for goodies for their lucky dips and bric-a-brac stalls. Alternatively, many charity shops will usually take bits and pieces of household furniture.

Bedroom ...
spring into action ...

Room for improvement

Do the contents of your wardrobe dominate this room? Do your clothes spring off the tightly packed hangers and sprawl onto the floor, causing chaos as they land on your eclectic selection of dusty shoes? Have you got suitcases full of 'stuff' under your bed? This room should be reserved for rest and play, and if you're feeling strung out by your strapless dresses, it's time for a serious shake up.

Some people have very strong issues surrounding their appearance; they feel they have to be in the right clothes and right shoes to fit in. However, if you embrace the prospect of living with less clutter by refusing to conform, you could save yourself thousands of pounds,

heaps of storage space, and still have a wardrobe full of fresh and exciting outfits.

A clothes swap party is a great way to get started. You'll find details in the 'Bedroom' Project box on page 79, together with an email to send to your friends.

If you've never been inside a charity shop except to drop off a donation, go and take a closer look; there are lots of bargains to be had. Charity shops vary wildly in the styles and quality of garments and shoes available. The posher the area the posher the garments, so if you're looking for something a little upmarket head to an affluent area. By spending your money this way, you can also contribute to worthy projects in your community that support local hospices or children's homes for example.

While we're on the subject of clothes, let's take them off! The *Durex Global Sex Survey* from 2005 showed that 43.57% of people in the UK have had sex using a vibrator. Maths really isn't my thing, but I had no trouble working out that from a population of just over 60 million people, deducting 25% for the under twenties left an excess of 19 million citizens. If we assume that the 'buzzy' devices were used by couples as opposed to solitarily, that gives around 10 million predominantly battery-operated devices. Many of these are shipped in from the Far East and, unsurprisingly, have planned obsolescence built in. Once the motor has burned out, or the battery has corroded the contacts, they end up in landfill. However, this needn't be the case as almost all of the components can be recycled. The UK company Love Honey have set up a 'Rabbit Amnesty', which allows anyone to return their old vibrators for recycling and, in return, they can buy a new rabbit for half price. The returned items are dismantled and recycled in an ecologically sound manner at a designated collection facility. Love Honey also donates £1 for each new rabbit sold through the scheme to the World Land Trust.

Change the habit, reduce the rubbish

• If you curl, tong or straighten your hair, consider changing your style for something that is far easier to manage and 'low maintenance'. Alternatively, get some funky headscarves and hats, hide your hair and give it a rest!

• If you must curl your hair, buy old-fashioned rollers or use the flexible coloured ones. After arranging the rollers, wrap a warm towel around your head while you are doing your make up. You'll look gorgeous and your beautiful mane will not have been subjected to harsh heat treatment that might dry it out, leaving it in need of a chemical treatment from another plastic bottle. Break the cycle, ladies!

• Take back to the shop any newly purchased shoes and clothes that you aren't going to wear. The charity shops are testimony to the fact that many of us buy things that we never wear. They can, by the way, charge more for items when the price tag is still attached.

• Crocs have become one of the biggest-selling shoes. They have a successful recycling scheme in America, and will soon be starting a similar scheme in the UK. Your old Crocs are recycled back into new shoes and donated to people in need around the world. Visit **www.solesunited.com** for details of the American project.

Cash in on reduction

• Never buy a one-wear outfit. Find a local hire shop in the Yellow Pages and borrow a posh frock instead, or simply borrow an outfit from a good friend or relative.

• Little electric fan heaters are expensive to run and are often dumped once people realise how much they cost. Wear an extra layer instead.

• Electric blankets are a waste of space during the many months they are not needed. Invest in a few hot water bottles for an inexpensive way of warming up your bed.

• If you have a special event coming up, use a hat hire company such as Felicity Hat Hire, who have outlets throughout the UK.

• To complete the look, visit Handbag Hire to borrow designer items at a fraction of their shop price (see Resources for details).

• Cheap and floppy foam pillows and duvets are a false economy. They break up after a few washes and are soon as flat as a supermodel's tummy. Invest in a good feather or duck-down version; they are easily washable and will last you for years and years.

• Soft glowing, low-energy LED lights are perfect for the whole house and particularly gorgeous in the bedroom. They will save you money and some come with lifetime guarantees on the bulbs/lamps. The old-style filament light bulbs or lamps are not recycled yet, but things could change, so check with your local recycling centre or municipal site to see what their protocol is.

• If you're still not convinced, or are unable to replace all your lamps with low-energy LED lights, why not use strings of fairy lights. They make sensual and moody bedroom lighting and you won't need to flick the mains lights at all.

• Get practical with your socks. Find a style you like and buy a job lot! This is a particularly great idea for children who are forever poking a toe through one or losing one. If getting a needle and thread out to sew just one is a nuisance, put your holey socks aside until you have a few and mend them all together.

• Stop buying trashy T-shirts that fall apart after a couple of washes. Go for fewer, more stylish organic cotton ones that show your commitment to the planet and will last you for much longer.

Reduce, re-use, recycle

• Get your sewing kit out and customise some of your outfits. It's great fun and easy to do. Find your nearest haberdashers and buy some patches, sequins, mirrors, crystals and braid to liven up your wardrobe.

• Revitalise a boring pair of jeans by sewing on a Velcro strip around the leg a couple of inches up from the hem, then get a colourful headscarf or piece of interesting material, cut it in half, sew the other side of Velcro to it and attach it to the trousers, giving you a trendy leg cuff!

You can get lots of different looks just by changing the attachment. This is also a great way to deal with young girls' trousers when they have a growth spurt.

• Once you've finished reading magazines and monthly journals, give them to a local retirement home, doctor's or dentist's surgery for others to enjoy.

• Following on from your wardrobe purge, and to prevent you filling it up again, put any old metal clothes hangers out with the tins for recycling and give the plastic ones to a local charity shop.

• If you go away for a short break, don't buy additional small bottles of toiletries to pack in your overnight bag. Instead, decant your cleanser, toner, moisturiser and other liquids into small empty containers.

• Borrow suitcases from a friend to take on holiday. Luggage is expensive and generally sits around for 48 weeks of the year doing nothing but collecting dust.

• Set aside posh bags and tissue paper from any lingerie you buy and re-use them for wrapping your own presents.

• Store old sheets in the garage or shed and use them to protect the furniture and carpet from paint splashes when you decorate.

• If your clothes have no life left in them, cut off any funky buttons and thread them onto strong thread or a leather band to make necklaces and bracelets, or tie them onto the ends of raffia or rough string to decorate gift labels or present wrapping.

• Old cotton T-shirts make great cloths to clean windows.

• Dried lemon peel is a natural moth deterrent. Just place it in your chest of drawers, or tie it into a handkerchief and hang in the wardrobe.

In the UK two million pairs of shoes are discarded every week.

Project box - Clothes swap

Throw a clothes swap party. Why? Perhaps the question you should be asking yourself is why not! It seems like a bizarre concept at first but, once you've tried it, you'll love it.

Start by having a thorough wardrobe purge and dig out all the outfits, shoes and boots that you haven't worn for ages. Whatever the reason for not wearing them, there's no reason for them to be cluttering up your life, mind or wardrobe, or potentially threatening to enter your bin. Many of your friends will be in the same situation, so use the email at the end of this section to organise your very first clothes swap party. Everyone that's coming should bring along a homemade small nibbly contribution, as the essence behind the event is sustainability and, of course, a bit of fun.

Once you've seen how easy it can be, you might consider doing it as a fundraiser in your school or local community. Punters could buy, say, five tickets for the privilege of swapping five items. They arrive with their bag of five items, which are checked to ensure they are of good quality and placed on an appropriate table, i.e. men's, women's or children's clothes, etc. The customer can then pick up five other items to replace them. Everybody wins; the charity of your choice can make a tidy profit, you have new (old) clothes in your wardrobe and nothing was thrown into landfill.

Any items of clothing that aren't suitable for your swap can be donated to a local charity shop or put in a clothes and textiles donation bin (as seen in many supermarket car parks) for Third World countries and disaster zones.

Wedding dresses are lovely keepsakes, but think hard about what you are going to do with it after the big day. If you cannot bear the sensible, money-saving option of renting outfits for you and your entourage, or if you have already tied the knot and are storing the most expensive garment you've ever bought in a bin bag, consider taking it to a seamstress and reworking it into a Christening or naming ceremony outfit for your children. Alternatively, sell it via your local small ads or donate it to a charity shop.

Use this email to entice your friends to your clothes swap party.

TO: friend@home.co.uk
FROM: Traceysmith@home.co.uk
SUBJECT: Clothes swap party invite!

Hi Friend

Just to let you know that I've finished reading *The Book of Rubbish Ideas* and it has inspired me to declutter the house and have a good clear out of our clothes. It has gone surprisingly well and I can actually see the floor of the children's wardrobes!

Nearly all of the clothes are in great condition. A lot of the kids' stuff is just too small for them now (and I know you've had your eye on a few of my tops) so I thought I'd throw a clothes swap party and invite a few of the girls round to my place next Wednesday, 1 p.m.

The rules are simple: have a good root through your cupboards and bring yourself and a friend over with your collective bags of clothes/shoes/boots and stuff, and a plate of nibbles too - nothing pre-packed please. This is going to be a celebration of sustainability while spending no money, changing our wardrobes and keeping fabric out of landfill!

If there's anything left over, I'll take it to the charity shop. And, if it all goes well, perhaps you could host another one in a few months time and I'll promise to let you borrow the book.

See you soon.

Bathroom ...
don't throw the towel in ...

Room for improvement

The bathroom is a prime area for amassing assorted disposables, including razors, electric toothbrushes, sanitary wear and nappies. However, if you open your mind to trying out a few longer-lasting alternatives, you will find yourself with a lighter bin bag and a heavier purse.

The vast range of epilators and electric shavers available these days can cut the wet shave out completely and eradicate the need for cans or tubs of shaving foam. A Sol-Shaver solar-powered razor will even eliminate the need for traditional power. Batteries required to power toothbrushes and other personal hygiene devices can create a waste problem that is easily rectifiable by switching to rechargeable versions. As with the solar shaver, you can also buy solar battery recharging devices (see Resources for details of solar equipment).

This room is also a magnet for bottles of assorted beauty products. There are a growing handful of ethical stores, like Who Cares? in Melksham, where you can concoct jars and tubs of your own beauty products using natural and organic ingredients. You are often encouraged to bring along your old bottles for a refill and are offered their recipe booklet to make up exciting and exotic potions, with an unusual range of ingredients you might not have in your kitchen cupboard (see Resources for further details). It's worth taking the time to seek out caring, ethical shops and suppliers in your area, as they might run a similar scheme for these and other products such as household cleaners, washing-up liquids and so on. *The Organic Directory*, edited by Clive Litchfield and published by Green Books, will be a big help, and your local independent health food stores will be a great place to start.

Disposable nappies account for an enormous amount of our landfill waste, with around eight million of them being thrown away every day in the UK. If you put your baby in disposables from birth through to the time they are able to go without one, it will cost you around £1,000. By using a 'real' nappy, you could save up to £600 on your first child alone. The Real Nappy Campaign helps you ditch the disposable habit. With a washable cotton nappy, you can use washable cotton wipes too and all you flush away is a biodegradable liner and the smelly contents. DEFRA, the Environment Agency and WRAP will be publishing an updated Nappy Lifecycle analysis later in 2008, which should make for interesting reading.

Of course, it's a very small leap to embrace the idea of washable sanitary wear and panty liners and there are many surprising options available here. The Mooncup is a silicone cup that sits very comfortably inside the body. It's very easy to fit, remove and keep clean and will not interfere with swimming or sport; it's a good idea to practise using it before you need to, but you'll soon get to grips with it. Many women note a considerable decrease in backache and stomach cramps while using it.

If you prefer an external option, look at cotton washable towels and panty liners. As with the nappies they also come in fabulous designs. Some of them 'popper-clip' onto a waterproof outer layer and their wings popper-clip around your underwear to hold them in place. They fold down to the size of a handkerchief in your bag and are very discrete. With fleecy 'dry' sides, they combine amazing comfort with great efficiency and just like the nappies they wash at low temperatures with your other washing. The cost savings and rubbish reduction from these two ideas alone are phenomenal (see pages 112–114).

Change the habit, reduce the rubbish

• Don't tip medicines down the toilet or clog up your bin with dangerous unused or out-of-date tablets. Simply take them to your nearest pharmacy for safe disposal.

• Say 'no' to throwaway wipes! Cut up a soft and threadbare old bath towel into small handy squares and use as daily make-up removal cloths. They will probably do fine for a couple of days, then just put into your next load of washing.

• Change your supplier of toiletries, skincare and make-up products for one that offers refills.

• Use an ethical high-street retailer for some of your bathroom beauty needs. They will often take back any empty bottles and canisters for recycling at the point of sale and give you incentives for doing so.

• Cut the end off your toothpaste tube. There's always a guaranteed blob of paste that will probably see you through the next week.

• When you buy your next bottle of perfume or aftershave, leave the mass of cellophane, cardboard box and inner padding at the till for the retailer to dispose of responsibly with their own commercial waste and slip the bottle into your re-usable cotton bag to bring home. If it's a gift for somebody else, do them a favour too, do the same and gift-wrap it in simple plain brown paper with a raffia tie and recycled tag.

Cash in on reduction

• Use an exfoliating hemp cloth instead of jars of scrubs that are empty all too soon. The cloth is only a few pounds, washable, re-usable and dries easily over the sink.

• Make up your own products using basic ingredients. There are many stores where you can buy glass bottles and pump dispensers too, so you can really 'go bespoke' in your bathroom.

• Use washable cotton nappies.

• Use re-usuable sanitary wear.

• Buy a rechargeable shaver and/or epilator and ditch the plastic options.

• Try doing without electric shavers at all. Just a good brush, ordinary soap and a traditional, re-usable blade will suffice.

• Cut out the cans of deodorant and try a crystal rock from an eco-retailer. They come in various shapes and sizes and have no chemicals or CFCs in them either. Simply wet the crystal and rub it under your arms for all day protection. They sell for about £9 and last for up to twelve months.

• Buy toiletries in the largest available sizes and top up smaller bottles of shampoo, conditioner, bath oils and so on.

• Install a wall-mounted push-button dispenser for shampoo and other products. You will waste far less by using a measured amount.
• Ask your hairdresser if you could buy the large salon versions of shampoo from them and remember these bottles are often more concentrated too; use a pump dispenser for economy and ease of use.
• Say goodbye to those dissolvable toilet products and pop an eco-toilet descaler into your cistern to prevent limescale build up (you may need two in very hard water areas).

> If everyone in England recycled one more plastic shampoo bottle, the energy saved would be enough for 46,000 households to watch a 32-inch plasma television for six hours a day for a whole year.

Reduce, re-use, recycle

• Instead of using lots of different bottles of bubble bath, try adding a handful of Dead Sea salt to the water to soften it. You can buy it in fairly large packs from your local health food shop, then store it in an attractive bowl. Finally, add a few drops of your favourite essential oil to the running water.
• Essential oils are amazing and so versatile. Add three drops of rose or lavender oil to the water for a relaxing bath, or grapefruit or lemon oil for a stimulating one. To help prevent nits add three drops of tea tree in the children's bath and add to their shampoo too.
• Throw empty tea-light holders from candle-lit baths into the metal recycling bin.
• Keep old toothbrushes for cleaning awkward to reach or tricky things like tap spouts, or put them in the children's art box to use as a creative paintbrush.
• Cut up old towels to use as hardwearing, household cleaning cloths.
• Have a bin in the bathroom for tissues, cotton wool, paper, light

cardboard, toilet roll innards and hair from the hairbrush and put it all in the compost bin.

• Don't wash as much! If you are hopping in the shower or bath every day, you are stripping your hair and skin of their natural oils and protection and then probably replacing them with a chemical cream or perfumed lotion.

• Trim the tips of old make-up brushes to give them a new lease of life.

• When you get to the end of your bars of soap, pass the little bits over a cheese grater and keep the shavings in a pretty jar; add a sprinkling to the bath to soften the water.

• If you have any cracked or leftover bathroom tiles, or if you've broken a mirror, use a tile cutter to break them into small chunks and use them to decorate a table top, plant pot or picture frame with a funky mosaic design.

• Think twice before you throw your clothes into the washing basket. If you've only had them on for a short time, hang them back up to air overnight and put them back in the wardrobe; they don't need a wash.

• Hang your bath towels out on the line after you've used them or pop them in the airing cupboard or over the banister to dry out. They certainly don't need washing after one use, or even several uses.

• Let the children loose with a pair of safety scissors and any spare plastic bubble bath bottles for a bit of cutting and sticking art club. They are great for making daleks with!

• Empty plastic squirty bottles make fabulous toys to play water games with in the garden.

• Cut the tops off the bottles to use in the garden as cloches over delicate plants, or as re-usable paintbrush cleaning vessels in the garage.

Project box - Smelly solutions

Now take note: essential oils crop up many times throughout this book. In fact they are roaring towards you right now in a subsequent paragraph and they are Mother Nature's pure and simple answer to a multitude of everyday problems.

They are great for both personal and household cleaning. Mixed together with other simple ingredients – many of which you will find in the kitchen – you can create safe and tasty treatments that make your skin feel beautiful from head to toe. You can also use them to add scent to plain massage oil and to treat minor ailments. Be sure to use only essential oils, not oils for room fragrances; and use as directed on the bottle.

Making your own will save you from buying packets and bottles of lotions, ointments and potions and even some medicinal aids too. You could even bottle or jar a few favourites as gifts for your friends. (See **www.bookofrubbishideas.co.uk** for more recipes.)

Here's my recipe for a luscious His and Hers body scrub for a sticky night in!

The combinations of this scrub are endless. This one is just a guide to kick things off, but it's really just a blend of a wet mixer, a dry exfoliate, and then anything you really enjoy the colour and smell of.

You'll need:
A wide bowl.
A mixing spoon.

To the bowl, add a tablespoon of one of the following ingredients:
Desiccated coconut.
Brown sugar.
Porridge oats.
Dead Sea salt.
Finely chopped nuts or seeds.

Add around a coffee cup per person full of liquid to blend. Use either:
Any carrier oil. (Almond or grape seed are lovely.)
Yoghurt with a spoonful of honey.
Mashed banana and a spoonful of honey.

Finally, add any one or two of the following if you wish:
A small handful of dried lavender.
A small handful of fresh rose petals.
A few drops of vanilla essence.
A little coarsely grated chocolate.
A couple of crushed strawberries or other berry fruits.

Simply stir the ingredients and apply gently, using slow, circular movements all over the body. Avoid contact with the face and eyes particularly. Then when you're ready, wash and rinse off.

Shaving tip

All you need as shaving soap is — soap. Use a good badger-hair brush and most soaps give you all the froth and help you need. No need for cans of foam etc, and no need for any electrical gadgets.

Sixty-two per cent of consumers recycle household items from their kitchen. Only 36% and 34% respectively remember to recycle items from their bathroom and bedroom. Forty-one per cent of people throw away recyclable items from the bathroom.

Garage/Shed ...
out of sight ...

Room for improvement

One of my most radical questions is, 'Do you actually need your car (or your second car) at all?' Your instant reaction will probably be, 'Arrrgh, yes,' but maybe you're not aware of the more cost-effective options open to you, so sit back and hear me out.

We have over 10,000,000 empty car seats whizzing up and down UK roads every day, often travelling ridiculously short journeys. They could so easily be filled up via a car-share scheme such as Liftshare. It was set up in 1997 with the initial aim of encouraging and enabling more people to share their car journeys. It claims to be the largest car-sharing company in the world, and is actively supporting a variety of sustainable transport measures.

Car sharing has many significant, positive environmental impacts. Over 40% of the members of Liftshare decide against buying a car, or even end up selling their car. Car usage could drop by as much as 50% if people found more  efficient ways of getting around. Walking, cycling, taxis, trains and other public transport methods can be brought into people's lives. Some of these methods may add a little time to a journey, but with determination and better planning they can be welcomed and enjoyed. (See **www.liftshare.org.uk** for full details.)

Many employers embrace the idea of car-share schemes to help their staff and are often willing to let you leave and arrive slightly 'off peak' to accommodate them. Be an eco-champion in your office and encourage them to start one up.

Zipcar and other car-club schemes provide another option. For a nominal membership fee you can hire out cars by the hour and return them to the same spot – an excellent option for the monthly shop. They have their own car-parking spaces, and each Zipcar replaces over fifteen privately owned vehicles. These schemes can change our urban landscape. Members of car-sharing clubs reportedly save up to £3,600 each year, and much of that is then spent locally, yet another upside for your immediate community. See **www.zipcar.co.uk** for full details of their scheme. There are now City Car Clubs in many UK cities.

Don't forget that no car in the garage means you'll have more storage space, which means you can buy in bulk.

Cars are often the last things to be found these days in garages. They are generally full of DIY equipment and assorted junk, which will probably end up in a black plastic bag destined for landfill.

Change the habit, reduce the rubbish

• Vehicle tyres last much longer if you inflate to the correct pressure. If you are not sure how to check this, go along to a tyre centre and ask somebody to show you.

• Drive your car 'gently' and it will make a considerable difference to the life of your tyres; you'll also save petrol.

• Get rid of half-used pots of paint and donate them to a local community repaint scheme. (See **www.communityrepaint.org.uk**.)

• Stop buying paper and plastic car fresheners. Fill the coin tray or the ashtray with little decorative stones and put a few drops of your favourite essential oil on them. It will last for ages and it's easy to top up or change.

• Used sump oil from the car should never be poured down the drain. This contaminates our waterways and can lead to blockage problems too. Drop it off at your local civic amenity site for recycling where it will be washed and turned back into useful oil.

• Take the time to measure up your DIY jobs properly and don't wildly overestimate your quantities. Stuff will probably end up sitting in a useless pile somewhere, providing food or sleeping quarters for rodents. Keep your receipts so that any surplus unopened stock can be taken back to the store. Any half-boxes of useful items can be given away on your local Freecycle group.

• If your vehicle needs a new part, ring around the local breakers' yards for a second-hand part before buying a brand new widget.

Cash in on reduction

• The garage or shed can end up being a graveyard for old battery-powered home-improvement tools. You can buy a new range of tools that use more efficient rechargeable lithium-ion batteries. They have more power, the tools are lighter, there is no self-discharge and memory effect so they are always ready for use, and they have a short charge time in comparison to conventional batteries. (See Resources for suppliers.)

• Speaking of batteries, did you realise that in the UK alone, 660 million batteries are bought each year, with the average household using around 21 batteries per annum. Of these, 95% are disposable and around 99% of them end up going straight to landfill sites at the end of their short but useful life. Disposable batteries have 32 times more impact on the environment than rechargeable alternatives and use 23 times more natural resources. Don't forget that you can soften your carbon footprint and save on utilities by harnessing the power of the sun and using a solar recharger.

• This is the place where we are likely to find other 'boys' toys', even if there are no boys in the house! Torches are a shining example, and they no longer need power from 'old-style' disposable batteries. Use lighting devices driven by solar power, wind-up energy or by being shaken. The LED lamps often come with lifetime guarantees.

• On the subject of illuminating ideas, providing light in remote sheds and garages devoid of electricity can be expensive. Consider fitting a solar-powered internal or external light.

• What do winter duvets, wedding dresses and garden furniture cushions have in common? We're often loath to throw them away, but they are difficult to store and take up precious storage space. Consider vacuum storage sacks that suck all the air out, preserving the items inside while considerably reducing their physical impact.

• Use easy-to-install, long-lasting, solar-powered lights as an inexpensive way to light up the path to your shed.

• When choosing a barbeque, don't be a lazy gas-guzzler. They are an increasingly expensive way to burn your burgers. Buy a charcoal-fired one Instead.

Reduce, re-use, recycle

• If you have worn your tyre down to the tread, don't leave it at the garage, but put it in the garden to use as a planter; they are particularly good for growing potatoes in!

• Hire heavy equipment that you plan to use only once or twice, such as large garden rotovators. It's a cheaper option, you can use the most up-to-date kit and you don't have to find storage space for it.

• Don't buy bottles of unnecessary cleaning fluids for your car. Add a few drops of washing-up liquid to the windscreen-washing reservoir for cleaning the front and back windows.

• Trim the ends off your paintbrushes to give them a new lease of life.

• Tangled string? Punch a hole in the lid of an old jam jar, insert a ball of string or garden twine and poke the end out of the top for a tangle-free storage solution.

• Use ripped bike inner tubes as draught excluders around loose-fitting or rattly doors.

• Don't store your bicycles outside. If you're stuck for floor space in your garage or shed, put up some simple wall brackets to hang your bikes on. If you must keep a bike outside, throw a waterproof tarpaulin over it for protection against the weather.

• Half-finished DIY projects often end up in the garage or shed. Be realistic, if you aren't going to get around to finishing them, call in a local tradesman. If money is tight, look around to see if your area has a LETs (Local Exchange Trading

scheme) in place and get yourself involved in a community-based mutual aid network. (See **www.letslinkuk.org** for further details.)

• One of the most unpleasant DIY jobs is the laying of new, or improving current, loft insulation. Existing insulation may contain fibreglass, which is dreadful stuff to work with. To find a solution to our resource crisis, we must buy 'new' recycled products too. Eco-Wool is made from 85% recycled plastic bottles while Thermafleece is made using wool from sheep that graze the upland areas of the UK. You can even get insulation made from shredded newspapers and recycled industrial denim, so shop around for kinder materials and stop wasting heat.

• Don't keep useless cables or electrical equipment that has long since lost its other half! Take advantage of the WEEE directive and deposit them at the store you buy a replacement from, or ensure they find their way to your civic amenity site for safe and necessary recycling.

An estimated 40 million litres of unused paint is stored and then disposed of in landfill. The DIY sector is responsible for the majority of this unused material. Research has shown that 25% of DIY sales remain unused, compared to 1.5% of trade sales.

Project box - Paint pool

According to the brilliant initiative Community RePaint Scheme, last year 400,000,000 litres of paint were sold in the UK to the trade and via retail outlets. It is estimated that approximately 56,000,000 litres went unused and were stored in homes or garages or simply tipped away or dumped — enough to fill twenty-two Olympic-sized swimming pools!

Community RePaint (CRP) provides a solution to this problem by providing an outlet for unwanted re-usable paint and helping local communities, individuals, charities and many voluntary projects at the same time. It's a national concern supported by ICI Dulux, funded from Biffa Waste via the landfill tax and the National Lottery. They collect good-quality, leftover paint and re-distribute it free of charge to eligible sectors of the community.

Why not appoint yourself a green champion in your street by organising an annual *en masse* collection of unused paint to donate to CRP. Use some friendly door knocking to pass the message on and don't forget to put a notice up at the local school, post office and corner shop with information about what you're organising.

If you are interested in donating paint or would like to request paint for a worthy project in your area, visit **www.communityrepaint.org.uk** for full details.

Garden / Allotment ... sowing the seeds of change ...

Room for improvement

This outside room holds the most potential for dramatically diminishing your rubbish and waste and, even if you don't have a garden, please don't skip past this section. There's more than one way to bath a baby, so read on.

I was brought up on a council estate in East London, and cultivating fruit and vegetables just wasn't my parents' thing. However, my grandparents grew runner beans and I have fond memories of sitting around the kitchen table with a cheese slicer helping to chop the bountiful yield ready for blanching and freezing. We had runner beans with everything for months, but that never put me off; they were delicious.

I once had no garden at all, just a concrete landing pad with a couple of raised beds full of weeds and the odd colourful bulb. I had great pleasure in ripping the rubbish out and planting vegetables instead. I had limited experience of growing anything, but learned by trying one or two new plants each year. I've had varying degrees of success with tomatoes, courgettes, aubergines, melon, squashes, curly kale, carrots, chillies, bell peppers, lettuce, Brussels sprouts, broad beans, cauliflower, cabbage, garlic, onions, leeks, broccoli and a few spuds, but not all at the same time!

For someone who didn't know the rear end of a pak choi from a fennel bulb, I haven't done badly and, between you and me, I paid little attention to the optimum sowing times and other 'rules'. Some worked, some didn't, but the experience taught me much in preparation for my efforts the following year.

I use no chemicals and have risen to the challenge of being inventive, catching crawly pests out with a mixture of organic methods and pure cunning. I remember the self-sufficiency guru John Seymour once said, 'Any fool can use chemicals.' How right he was. When my children go out to find something for tea and pick 'one for me, one for the basket', I don't have to worry that it hasn't been washed.

If you bring to your table a few items you have reared organically, the sense of joy and achievement is memorable, even if you only harvest enough for one meal. If you only focused on one crop, the versatile tomato for example, in a raised bed, or a hanging basket, or a windowsill box, or a patio pot on your balcony, you could cultivate a hefty yield. If you don't have any outside space, why not consider putting your name down for an allotment.

If you are going to grow seeds, you will need something nutritious to grow them in. This is where your own compost comes in handy. If you think compost is smelly and difficult to make, think again; it couldn't be

easier, and I will show you the basics for making compost, using a compost bin, a wormery and a Bokashi bin.

Composting can transform many of the items you currently throw in your bin in a short space of time, turning them into what's often known as 'black gold' or compost.

Let's take the composting bin first. You can put in anything that was living, with the exception of meat, dairy and cooked food. These would biodegrade in time, but they attract vermin. More advanced composters, called Green Cones, are buried into the ground and allow for some of these other items, including chicken carcasses, but breakdown takes longer and they have to be installed correctly.

The standard composter — a plastic or wooden bin — needs to be sited on flat earth, so that the worms and other creatures can enter. Add a variety and balance of wet/green things, dry/brown things, and ideally some activators to get it all going.

Typical green things include grass cuttings, nettles, raw vegetable peelings, fruit waste, tea bags, tea leaves and coffee grounds, small prunings, rabbit or other vegetarian pet manure and its bedding, and herbivore manure from cows and horses.

Typical brown things include cardboard (cereal packets, egg boxes etc), waste paper, non-glossy junk mail, toilet-roll innards, newspaper, old bedding plants, dead flowers, sawdust, wood shavings and leaves.

You can also add hair, cotton wool, used tissues, the contents of your vacuum cleaner, toe-nail clippings, wool and any other natural fibres (ideally cut up), crushed eggshells, paper bags and wood ash from the grate.

Activators speed up the decomposition process and one of the best is male urine, which is rich in nitrogen, so get the man in your life to do his bit for the environment once a week if possible.

If you are concerned that there will not be enough room for all your food and rubbish in the bin, you'll be surprised how quickly it squashes down and turns into crumbly earth. However, when buying one, ask for one that can deal with the number of people in your household. Compost bins that have doors at the bottom are ideal for scooping out the fully processed stuff; all you have to do is keep adding raw ingredients.

Wormeries work in a similar way, but because they come in sealed units you don't need to site them on the earth. They can sit in an apartment lobby, or in some outside space in your office or school where they will consume your colleagues' tea bags and fruit leftovers. Your new wormery will come with a bag or two of live worms, often tiger worms, who are voracious creatures that eat more than their entire body weight in food every day. They turn your rubbish into vermicast (worm wee), which can be drawn off from a tap that should be sited at the bottom of the wormery unit itself. This liquid is the most incredible natural fertiliser and wonderful to use on your seedlings and new plants.

Bokashi bins put the icing on the cake of composting. They complement the composters and wormeries you are already using and allow you to turn all your kitchen waste into nutrient-rich compost, including meat, fish, dairy products and cooked food. Place all the waste into the airtight container and sprinkle with a handful of Bokashi. This is a bran-based material made with a culture of micro-organisms, which begins to compost the waste. Inspired by the Japanese, the EM-Powered Bokashi performs the first stages of decomposition using the active micro-organisms. After a few days the contents can be safely transferred from the Bokashi bin into your composting bin or wormery.

For information on all of the above, visit the experts at **www. originalorganics.co.uk**. They will be able to supply you with everything you need to get started. There are lots of great websites that will give you the ins and outs of composting.

Now that we've dealt with all the food, what else is left? If you have a dog, one of the biggest problems is dog poo. You cannot put their deposits in your composter, but you can buy a special composter to deal with it (see Resources).

Change the habit, reduce the rubbish

• Choose the most convenient method of converting most of your household rubbish into compost and use a bin, a wormery or a Bokashi bin.

• Compost your pet poo in a special bin.

• Don't buy plastic disposable plates and cutlery for a barbeque; if you don't have enough utensils to serve food and drink in, ask your guests to bring their own. Don't use polystyrene cups. We currently produce around 24,000 tonnes of these each year and they really are bad for the environment.

• It is safe to compost newspapers as long as they are ripped or shredded, but bear in mind that it's better to put your paper into the waste paper collection for your recycling truck to collect for pulping.

• If you are going to grow a few things in the garden, reserve a patch of land nearby and let it grow wild. Plant a few highly scented flowers there. The bees it attracts are about to become your new best friends.

• When you sow your seeds, always sow more than you need. If you end up with everything blooming, share your abundance with a friend, or cook, freeze or preserve it.

Cash in on reduction

• Growing fruit and vegetables is great, but beware the slugs! Copper pipe can be laid around the edge of your beds and commercial slug rings are available. A 15 cm length of plastic pipe with copper foil tape around it can be used to protect individual plants.

• Don't buy bottles of toxic bug repellents; instead plant out Thai lemongrass plants on your patio to deter mosquitoes. Break off a stem, rub it between your palms and apply to exposed skin to give direct, natural protection.

• Buy a water butt to collect your run-off water and use it on your tender new plants and seedlings. For something funky, see **www.waterbutts.com**.

• Get a bath siphon and drain out your used water to use in the garden.

Reduce, re-use, recycle

• Grapefruit and other citrus peels, used coffee grounds and mixtures of eggshells and sharp sand or crushed mussel shells can be used in gardens around plants to repel slugs and snails.

• Tie old or freebie CDs and DVDs to sticks with string to frighten the birds away from your crops.

• Get involved in the garden swap shop and swap and share your garden abundance! See **www.gardenswapshop.co.uk**.

• Swap your abundant crops with a friend who grows something different. You might end up with a bag of something delicious that you don't grow.

• Use old egg boxes for seed trays and transfer seedlings into old yoghurt pots when they are ready to thin out.

• If you're looking for a composting bin, make your local council the first port of call. They want to get you on the green track and often offer bins at a heavily subsidised rate.

• Use old bubble wrap from ripped jiffy bags to line plant pots and protect them from the cold, but be sure to make drainage holes at the bottom of the pot.

• Use an old net curtain as a protector around any tender plants that the birds might be partial to nibbling. Stake it down into the ground on wooden pegs, or tie it loosely over your fruit bushes, allowing room for the bees to get in.

• Use a cardboard box to bury your much-loved tiny pets in the garden.

• Be an eco-champion in your office. Start a composting or wormery scheme for tea bags, coffee grinds and fruit waste.

• If you are already a converted composter and know the benefits, why not consider starting a community heap in your area for those who don't have a garden? If there is a piece of scrub land nearby, it might be worth making enquiries to see if you could put it to better use.

Project box - Recycling more batteries

The plight of the battery hen has received much attention in recent months. Hugh Fearnley-Whittingstall, Jamie Oliver and other household names have exposed the harsh truth about their short lives and their grim living conditions. Hugh also launched the Chicken Out campaign to promote free-range meat.

Some years ago, I liberated ten ex-battery hens. They cost 60 pence each and barely had a feather between them. Their claws were overgrown to the point where they couldn't stand properly, their beaks had been blunted (de-beaked) to stop them pecking each other and, although they were a year old, they were about the size of my other three-month-old chickens.

When they stood on terra firma for the first time they instinctively knew how to scratch the ground looking for grubs. I watched them take their first taste of grass and leaves; one stood by my blackberry bush and gorged itself until it was almost sick.

They were terrified little creatures that didn't know what night-time was, as they had been used to almost 24 hours of fluorescent light. At dusk, they stood outside the chicken house as the other girls piled in and found their perches for the night. I had to lift the battery birds into their safe and comfortable beds, and they ran their beaks up and down the wire on the doors in fear that they wouldn't be allowed out again.

With basic food, warmth, love and care they were soon nursed back to full health, with a complete set of stunning plumage. In no time at all, they were laying and producing the most amazing free-range eggs I'd ever tasted.

You could keep some chickens in a little space in your garden. Your kitchen scraps and vegetable peelings would make excellent food for them and you will get free eggs in return. For help and advice on sourcing some chickens and suitable housing, please contact the Battery Hen Welfare Trust (see Resources).

Garden centres are becoming increasingly popular as more people try their hand at growing their own vegetables. They are also the source of much plastic packaging in the form of pots and trays. If your local garden centre does not provide recycling facilities, maybe you could convince them to start.

TO: manager@gardencentre.co.uk
FROM: Traceysmith@home.co.uk
SUBJECT: Garden rubbish reduction

Dear Garden Centre Manager

A recent news report on the BBC informed me that many garden centres up and down the country are starting up their own recycling schemes. They enable their customers to deposit used plastic plant pots and seed trays into a collection box at the front of the store.

Apparently, some 500 million unwanted pots are thrown away every year in the UK and, no doubt, a great many of these end up in our landfill sites.

As a keen and responsible gardener, I take great pride in being kind to the countryside and I'm urging you to set up your own scheme and to provide a receptacle for your customers' old containers.

If more plant pots were recycled in this country, we could encourage re-use and produce more recyclate for reprocessing into new pots and other things. There could be better opportunities within the UK for manufacturing with plastic recyclate, reducing the amount of product we currently have to ship abroad for this process. Everybody benefits!

However, in the first instance, we need to encourage change by offering convenient, local facilities for plastic plant pot recycling.

I am very hopeful that you will support this initiative. I have copied my email to our local radio station and our local MP, who I'm sure will be very interested in following the story. This would, of course, give worthy exposure to garden centres who implement a recycling scheme and who were clearly seen to be taking their corporate social responsibility seriously.

I look forward to hearing from you soon.

cc: Your local radio station
cc: Your local Member of Parliament

PART 3

What others are doing

Case studies ... be inspired ...

I'm not a girl easily swayed by persuasive advertising. I am far more likely to buy something, or change a habit, because of a recommendation from a trusted friend, especially if it comes from a fellow downshifter. If you discover something new and amazing, there is a satisfying buzz as you pass the message on to a chum.

The following case studies come from people who have an important message to spread. They are trusted sources and I respect and value their opinions. I hope you find something from their experiences to pass on to another good friend.

1. Living a sustainable life

We can look to governments, politicians and neighbouring countries for suggestions on ways we can cut down our waste, but ask yourself honestly, how likely is it that their ideas are going to affect your everyday life and behaviour? We are far more likely to take up a great tip passed on by a friend while we are waiting for the kids to come out of school, or from a chatty chum in the queue to buy a lottery ticket.

We 'everyday Joes' have an incredible ability to provoke change and to encourage it in others. The case study of this lady shows a perfect example.

Debra Rawling, 41, Jeremy, 39, and their daughter Emily live a sustainable life in Northamptonshire.

I usually have a very happy demeanour and the patience of a saint, but the amount of packaging and over-wrapping of the things I buy at the shops makes me really cross. Some items are even double-wrapped, such as multi-packs of canned foods and yoghurts. The difficulty of opening some of that packaging is irritating enough, but then having to throw it in the bin seems completely stupid. Easter eggs are a classic example of huge amounts of packaging being used for something that's mostly filled with air. Many medicines and beauty products are contained in an extra outer box. I know this is for ease of storage and transportation but I really don't think it's necessary.

I often shop with my daughter, and at our local supermarket we have to weigh and price up our own fruit and vegetables. Emily rarely puts them into a separate plastic bag. Wherever possible, she puts the price sticker directly onto the item itself, as in most cases nature has already done a good job of packaging it. We also take our own re-usable bags to places we visit and decline offers of plastic bags.

I love my simple life and adore my family and surroundings. I feel so lucky to live on a beautiful planet and don't want to spoil it by turning it into a huge rubbish tip.

Emily is only twelve years old, but she's very concerned about the planet and the life it sustains. She has definitely been the inspiration in our family to take stock of our habits and make positive changes. She is always coming up with bright ideas for new ways to reduce our family rubbish.

When Emily was about three years old, she asked us where all our rubbish went. I explained to her that it went to landfill and she replied, "The world is going to become a big rubbish dump then!" This comment from such a young child was a real eye-opener for us and I think it was the point where my husband and I really started being more conscious of our environment.

Living sustainably has become second nature now, and re-using and recycling is part of the way we live. However, there are important questions we should all be asking ourselves if we are going to get this country standing on a pair of proper green feet!

Consumer pressure is everywhere these days, but we shouldn't bow down to it. Don't keep buying stuff you don't need — that's a pretty good place to start. Do we really need to bring more items into our homes, especially when other perfectly functional items then have to be thrown away? We should think more about whether we actually need something.

When something stops working, Jeremy will always try to repair it, rather than throw it away and buy new. We do not own a toaster because we have a grill and don't need two items that do the same thing! We do not waste food; we menu plan and buy only what we need. We make a huge effort to use everything up; virtually nothing ends up in the bin and we compost all the organic matter to use on our vegetable patch.

It's easy to make recycling central to everything you do. In the kitchen, we've recycled an old plastic unit on wheels with two drawers and open space underneath and use it to hold all our paper, glass, foil, cans and plastics, ready to go outside on collection day. There is also space for our compost tub on top, so there is no fuss when it's time to 'do the bins'.

I'm a busy mum, but I make the time to do the things I know are important. There is no excuse for not recycling when the truck stops right outside your house. Sprawling landfill sites have devastating effects on the delicate eco-balance and wildlife, and it is we who 'create' them with our ridiculous waste.

What will life on this planet be like for our grandchildren's children? We have not been consuming at this level for long and we are in a mess already; it cannot continue. We all need to recycle much more and consume less, re-using what we have to minimise the rubbish we produce.

The Native American people lived sustainably on the earth by keeping in balance with the environment. They believed that what they do today impacts on the following seven generations. If we all adopted this approach to living our lives, the planet would be in safe hands. Let's stop thinking of ourselves and consider the future of our beautiful planet and all its life-forms.

2. The quest for a size zero 'waste line'

There are some people that just love to sink their teeth into a problem and shake it about until they get a satisfactory result. This case study highlights the plight of a regular mum turned rubbish reduction campaigner, who really put herself to the test on Zero Waste Week. She rose to the challenge with style and determination and whooped up a frenzy of interest from readers who followed her progress via an online diary called The Rubbish Diet.

Freelance writer and rubbish activist Karen Cannard, 39, lives with husband Adrian and their two children (aged three and under) on a modern housing development in Bury St Edmunds, Suffolk.

My interest in rubbish started when I became aware of the impact it was having on my personal environment at home. Everyday I was dealing with what felt like masses of unnecessary packaging. Sorting it all out for recycling — when it could be recycled — was a time-consuming job.

The biggest problem in our household was the black bin, which lurked in the kitchen as a constant reminder of all the rubbish our family sent to landfill. It filled up almost as soon as it had been emptied and we were throwing out nearly three huge bags every couple of weeks. The smells were horrid, particularly during the summer months, and the bin was so heavy to empty that I would often leave the job for my husband to do on his return from work; it wasn't the best greeting he could have had. It was, 'Honey, I'm home,' followed by, 'That's great darling, now empty the bin.'

I first became aware of the environmental impact of landfill through the waste reduction campaign group WRAP, in particular their Love Food Hate Waste campaign. The campaign helped me to recognise the levels of food waste that go to landfill and the consequences this is having on the environment. I tried to reduce waste by keeping a close eye on the amount of food that we bought, cooked and served at family mealtimes, making sure that I only cooked just enough or that I could use or store any leftovers. It had some impact on our bin, but I found it very difficult when catering for the tastes of four different people.

One Christmas I made a greater effort to buy food with minimal packaging, and that New Year I made a resolution to reduce our family's waste even further. I had no idea what I was going to do about it or indeed where this new resolve would lead.

A friend sent me an email about Zero Waste Week, a campaign run by our local council, calling for local residents to sign up for the challenge of reducing their household waste. The council was offering lots of hints and tips to participants and, as I needed all the help I could get, I promptly registered.

The council emailed a set of questions on our household rubbish and I replied telling them about my plans to reduce. I received a speedy response and was asked if they could feature me in their corporate magazine with a photograph of my bin! That turning point led to the creation of The Rubbish Diet. My challenge was going to be read by lots of people I knew and I had to take it seriously. It was as if I had publicly declared that I was going on a diet, and I didn't want anyone to see me fail. I enjoy writing and I knew that if I set up a blog (**www.therubbishdiet.co.uk**) it would keep me motivated and provide a forum for me to ponder the stresses and successes of my new challenge. Soon I started receiving advice from people as far away as America, and there were many messages of encouragement. I made small and simple changes to our habits, which enabled me to achieve almost zero waste.

When I received a request to record a column for Radio 4's Woman's Hour, which was to broadcast every day to coincide with our local Zero Waste Week, I realised the significance of my action. I was an average woman, in an average household, generating very little waste, then suddenly there was a widespread interest in what I was doing and how I achieved it. I could see that The Rubbish Diet clearly had a wider role to play in society. If it can help people improve their personal environments as well as the environment at large, I know my efforts have been worthwhile.

My top tips for a size zero bin are: examine what you buy and what you waste, avoid packaging by buying loose and local, re-use to avoid those disposable pitfalls, learn to cook with leftovers, get a composter or wormery, and recycle.

To keep up to date with my latest adventures in the world of waste, log onto **www.therubbishdiet.co.uk**.

3. Local authority action

When it comes to waste minimalisation and forward-thinking local authorities, St Edmundsbury has earned a good few Brownie points and has a place with some of the best. They were one of the first in the country to roll out the Zero Waste Week campaign, an initiative that brings together the community, local government and local media to create long-lasting change.

Kate McFarland is the Waste Development Officer for St Edmundsbury Borough Council in Suffolk. She lives with husband Tim, rabbit (Dave) and a chicken (Brian) just outside the market town of Wymondham in Norfolk. Kate is a passionate advocate for waste reduction and clearly an asset to this East Anglian local authority.

I started working for St Edmundsbury in September 2006. I have helped introduce a kerbside battery collection, the first of its kind in Europe (for all types of household batteries), and small WEEE (waste electrical and electronic equipment) collection, collecting small electrical household items. Both of these trials are funded by WRAP.

Zero Waste Week is the first project I have initiated. The campaign starts on 10 March each year and aims to give people simple ideas on how to reduce their waste. There are hundreds of residents signed up, all committing to reduce their rubbish. It has created a storm of interest with the local media, with coverage in local papers, television and radio. Businesses are also supporting us, as they recognise waste reduction is key to keeping their customers happy.

Waste minimisation is a real passion for me. I'm always looking for ways to save throwing things out. There are so many common-sense ways of reducing waste that don't need a huge amount of effort to implement, and if you can save a few pennies from these ideas too, even better.

St Edmundsbury has always been a forward-thinking and innovative council; they were the first in the UK to break the 50% recycling rate. This, of course, reflects the commitment of the people in the borough. The next step was to help them reduce the overall amount of waste created, and the response from members of the public has been immense. People don't want to have to deal with the volumes of waste they are creating, so the idea of stopping it coming through the door has been appealing.

The results of the Zero Waste campaign will be available on **www.stedmundsbury.gov.uk/zerowaste**, together with a list of top tips for reducing waste in your home.

The Zero Waste Week campaign has definitely raised public awareness of how wasteful our society has become, and we have helped steer many people towards waste reduction. A few simple changes can make a big difference; this project is proof and we hope other authorities will be inspired to run similar waste reduction models in their areas. We hope that we will soon be able to roll Zero Waste Week out across Suffolk.

My personal interest in waste started when I took a waste management course at UEA in Norwich as part of my Natural Science degree. It was only a short unit, but it really inspired me. My catch phrase soon became 're-use is better than recycle' and I started to drive my friends mad with my new-found knowledge on waste.

One of the first modules was exploring everyday appliances in one of our practical sessions; they were clearly designed not to be repaired easily. This meant that if anything went wrong it was cheaper to buy a new kettle, toaster or telephone than get it fixed. I thought this was wrong, and wondered how the manufacturers got away with it. Recently, with the introduction of the WEEE directive, they now have to take responsibility for the recycling and disposal of their products at the end of their life. This will certainly reduce the number of white goods that end up in landfill sites.

After I left university, I worked for a short time at Breckland Council in Norfolk, where I took a keen interest in other waste issues such as littering and fly-tipping. As a child I was always taught never to litter, a strong principle that has stayed with me to this day.

4. To inform and encourage a sustainable lifestyle

If you take it at face value, this is a bizarre process; we are printing on paper that originates from trees, to spread the message about not cutting down the trees and being kinder to the environment. Responsible publishers are taking this very seriously by using recycled or sustainably sourced stock, and others even offer online alternatives, turning away from the printed option entirely. Perhaps the days of popping to our friendly newsagents for a copy of the *Daily Truth* are numbered. This case study looks at a publisher with a foot in both camps.

Green mover and shaker Marc De'Ath, 28, is an ethical publisher, creator of *Sustained Magazine* and founder of the Grass Roots Alliance. He is based in Colchester.

Some years ago, I found myself pondering the fact that, as a nation, we were beginning to question our behaviour, our lifestyles and ourselves. It seemed clear to me that we were starting to make the connection between our actions and their impact on the planet, its environment and its people.

As I embarked on my personal journey towards living more sustainably, I began to feel quite overwhelmed with it all, depressed even. With so many choices at every juncture, I found it increasingly hard to decide what was actually ethically right and wrong, and I even struggled to decide which of the many green issues I was being confronted with were the most important.

I found myself feeling guilty for choosing one cause over another, and I thought to myself at the time, it isn't just me, there genuinely is a lack of public knowledge. It had left me unable to make a truly informed decision, and I was certain there must be other people out there feeling the same way.

This fuelled my energy and an overwhelming passion to create *Sustained Magazine*, a free publication aimed at the general public. I went about establishing a community with a group of ordinary people, tackling important everyday questions and covering topics that affected us all. It was a fantastic vehicle to talk about our daily eco-quandaries.

I decided to publish it both online and in print, giving people who were concerned about how much rubbish the printing industry creates, and the vast amounts of paper in circulation, a completely guilt-free read via the Internet.

The Internet also gave our readers a chance to share their own experiences and comment on our articles. The digital age has provided us with an alternative to printed media, but not a replacement. We are a long way from replacing paper as the main base for distributing our reading material, but it's great to offer choice. We use 100% recycled paper from sustainable sources for the magazine.

We recently ran a feature in conjunction with London Bio Packaging, tackling the topical issue of packaging. We must look at these subjects from all sides to get a clear and informed viewpoint. It provided a detailed breakdown of the many different types, with their pros and cons. The feature also included helpful tips and encouragement for our readers to reduce their packaging-based rubbish.

With paper and card making up about a fifth of the contents of the typical household dustbin, we need to know more about how to dispose of it in a responsible manner. As members of the public, we have the power to refuse or return items that unnecessarily use virgin papers and, as publishers, we need to recycle 100% of our waste and print on recycled or sustainably sourced material wherever possible.

I believe we all need to live a more ethical existence; sustainability naturally follows. A sustainable lifestyle is one that limits our impact on a local, national and international scale. Couple this idea with good ethics and we can make a difference, not simply by minimising our damage, but by improving our community on every level. The age of using virgin papers to wipe our backsides is over.

5. Re-usable nappies

Having your own baby or small child is not a prerequisite for reading this important case study. You might know somebody that has given birth recently, or is due to sometime soon. This could help you wrap up the baby shower presents in one hit. If you don't know anybody that fits that description, and you're a woman, there's news here for you too.

> Mark and Helen, young thirty-somethings, live with their sheep, lots of cats, chickens, dogs and the children who inspired it all, in a peaceful corner of Wales, and provide employment for several local people in a rural economy.

After the arrival of Hazel, our firstborn, we were horrified at the sheer bulk of waste produced by disposable nappies, not to mention the terrible stench they created in the rubbish bin. We decided to switch to re-usables when she was just a few months old. When our second child Joe came along, both of the children were in nappies and I found myself changing an awful lot of bottoms! On the plus side, Mark and I became dab hands at the whole thing; our bin was no longer smelly and we had halved our rubbish overnight!

We tried and tested many different nappies and got on surprisingly well with them. It was a pleasure to spread the word about how easy and reliable they were to use; they were a far cry from the old terry nappies and the designs were quite funky. My health visitor used to say, 'What have we got on today then?' when the children went for their health checks, and she was referring to their nappies, not their outfits! I took quite a bit of unpaid maternity leave and only worked part time, so we started really to appreciate the financial savings we made with our washable nappies.

I decided to set up a website to encourage people to give them a try. At the time, Mark had a huge commute to work as a secondary-school teacher, but he also got involved and wrote the web pages. Then, after much deliberation, we decided to take the plunge and be a real nappy retailer! Mark gave up work and BabyKind went live in April 2004, but we had no idea how many parents would convert to washable nappies.

While knee deep in research one day, washable feminine hygiene products and menstrual cups came to my attention. Initially I thought the idea preposterous; those things were only for 'hard-core greenies'! Then some of our customers started asking for washable sanitary products, and I realised the impact on landfill from these items is also enormous, especially when you consider that a woman usually menstruates for thirty years or more. It was a logical step forward, so we started stocking a selection and sales took off quickly. People raved about how comfortable the products were, but I still hadn't tried them myself!

Back then, we ran the business from our house. The office was half of our living room and all the stock was in our loft, which was converted into two small rooms. One day I had to raid our stock box! What a surprise I had; far from being leaky and uncomfortable, as I had somehow imagined they would be, they were soft, slim and did a great job. I washed them with the nappies and they came up as good as new. I stored the pads dry in a bucket, gave them a cold rinse in the machine and then a wash at 60 degrees. They last for years, save you a fortune, and have zero impact on landfill.

The best way to change people's opinions of washable nappies and sanitary protection is by letting them try them. You can tell people how great they are and they sound vaguely interested, but will then give you a list of reasons why they can't/don't use them, such as they'll leak, they're too expensive, they're hard work to wash, they're bulky, their lifestyles are too busy to fit in any extra work, or the baby has sensitive skin!

We decided to allow people to try the nappies and offered a 70% no-quibble refund. Initially we thought this would be a big risk for the business, but actually very few are sent back, maybe one in a thousand. We also hire out a nappy kit and have a dedicated advisor to help with choices and support, and we've set up an agents' scheme. We have around a hundred agents all over the UK, all of them parents who want to spread the re-usable word and reduce our impact on landfill. For more details, please visit our website, **www.babykind.co.uk**.

6. Work-at-home recycling

The work-at-home market is getting bigger by the day. Mums and dads everywhere are downshifting and ditching their commute to the big smoke to be part of a new wave of self-employed souls who've shrugged off their corporate stresses.

But I think you have to work harder than ever before when you turn this dream into a reality. You have to be savvy with your spending, and a sensible work-at-home parent makes every resource count and keeps her eco-footprint to a minimum.

This case study looks at a switched-on lady who has found the perfect blend of work and play.

Natalie Yeates lives with her husband Jools and their two boys in the Vale of Pewsey.

Life in the Yeates house is green and simple and, as a family, we take the whole subject of sustainable living pretty seriously. I'm part of a rapidly expanding wave of work-at-home parents and pride myself on having a good handle on the work/life balance.

My background is in interior design and, when the children came along, I decided to trade in the long hours in exchange for more time to be at home being mum. This allowed me to put delicious home-cooked food on the table every day, but I also wanted to have a creative outlet and bring in an income too, so I created Seventh Moon Wedding Design (see **www.seventhmoonweddingdesign.co.uk**). It is a small wedding stationery and accessories business and I make everything by hand, using lots of recycled materials throughout my ranges. My aunt brings over lots of old card and packing materials from her business too, for me to re-use. It's important to me that I offer an eco-friendly range.

Weddings are incredibly opulent affairs, with the average cost being around £14,000, and there's a shocking amount of physical rubbish created during the big day. An enormous proportion of it ends up in landfill. It's inspirational to see many brides taking a more thoughtful approach whilst giving consideration to the clearing up! For example, some use plantable seed sticks (recycled card with seeds impregnated in them, ready to be put straight into the ground to grow) or wooden name tags as wedding favours and re-usable place markers; some prefer silk flower arrangements, pine cones, twigs and even seasonal fruit and vegetable displays to decorate the tables, as opposed to fresh flowers that are dead within a few days. Some use edible table decorations like their favourite sweets for place markers or table centres. Most venues these days stipulate that biodegradable confetti must be used within their grounds; this can be made from flower petals or water-soluble rice paper. Really ethical brides have gift lists made up from charitable organisations such as Oxfam. Their guests can buy 'a bit of a school, or chickens and other livestock' for remote villages in Africa, as opposed to traditional wedding gifts like the good old toaster. Using sustainable methods to celebrate your wedding day isn't seen as scrimping or being cranky; in fact, it's becoming positively cool!

Working from home lets me have an eco-thoughtful lifestyle, both for my family and business life, and I often take rubbish from one side and turn it into something useful for the other. The upshot is that we have a pretty slim bin.

I use recycled and standard materials in my work; I'm a real squirrel and waste nothing, I use recycled items for my business, such as jiffy bags and cardboard cut from boxes that I use as a stiffener in my envelopes. Table plans are wrapped in old boxes and card, along with recycled bubble wrap or biodegradable corn chips. My favourite wrapping material is good old brown paper, which is strong, inexpensive, and completely re-usable at the other end.

Small businesses can be extremely wasteful, with excessive printing being one of the main culprits. Most of my 'waste paper' is shredded and used for packaging or as glue sheets or to start off our wood-burning fire; if ever I get too much, I pack it off to the recycling centre.

I work from all over my house, usually where the best light comes through the windows – why have a light on if the sun is shining – and part of my 'office' is in the hallway. It has no heating and a door sausage made from recycled pillows stops the draughts.

I do a little voluntary work in our local community shop and my colleagues are also very sustainability minded. There's a great feeling of community spirit there and we are all thrifty and resourceful. The shelving is made from recycled materials and other second-hand items and we only use plastic carrier bags donated by the villagers. To be honest, most customers tend to bring their own re-usable ones.

My life revolves around recycling and re-using, saving on packaging, buying locally, having a veggie box delivered – we return the boxes of course – visiting my local farm butchers and buying in my village shop. It's a thrifty, eco-thoughtful, creative, happy, family-first, work lifestyle that benefits everyone and the planet, and I wouldn't swap it for the world.

7. Recycling with Freecycle

The Americans get some pretty bad press when it comes to being green, and it's about time we stopped tarring everyone with the same Bush, sorry, brush.

Some bright spark by the name of Deron Beal certainly broke that mould when he thought up the concept of Freecycle. On 1 May 2003, he sent out the first email announcing the Freecycle Network™ to about forty friends and a handful of non-profit organisations in Tucson, Arizona.

The premise of the idea was to give away perfectly good household items to people that needed them, thus keeping them out of landfill. The concept has since spread to 75 countries and there are millions of members all around the world.

This next case study is about the organiser of a vibrant group based in the southwest.

> The tenacious Paul Walker, 45, runs the very successful Chard and Beaminster group of Freecycle. He lives with his partner Judith, and sons Aaron and Connor.

I reckon there are about ten years of landfill space remaining in the southwest. Whatever the figure is, once it's filled up, and if we continue to consume goods at the current rate, what are we going to do? One solution could be incinerators, but they come at an even bigger cost to the environment. There are far too many re-usable items sent to landfill, and this is something we have to address before it gets any worse. I'm doing my bit by running a Freecycle group. If you've never heard of it before, the global empire comprises 4,290 groups, with over four million members in local community groups all around the world. The creators simply describe it as a 'grassroots and entirely non-profit' movement for people who give and get stuff for free.

It's a big concept to comprehend at first, but here's a typical example to help you get your head around it. If you are changing your dining chairs for a different style and your old ones still have lots of life left in them, you could put a posting up on your group offering the chairs to whoever needs them. Other members read through the daily postings and drop you a note if they are interested in collecting them.

Once you've got into the swing of it and given some things away, you can post 'Wanted' notes too. It's a brilliant way of keeping things, particularly large items such as furniture, out of landfill sites. Examples of things that have been Freecycled are plates, greenhouses, plants, phones, pots and pans, tables, beds, clothes, washing machines, books, even cars and just about everything in between. Membership is free and you don't have to be a techie to make the postings either; it's really easy.

I'd been aware of the rapidly expanding Freecycle movement for quite some time before I started up my group in Chard and Beaminster. There was a lady in a neighbouring group who I used to natter to and, after many discussions, I started thinking about starting a group near me. I'd already started a 'Man & Van' service locally, following a long-term illness that left me unable to return to the construction industry. Being a man who likes to keep himself busy, I realised this was something I could do in my spare time, and I'd also be able to indulge my passion for reducing rubbish.

I carried on with life as usual and then, one day, I was doing a small job for somebody and dropped a few bits off at the local recycling centre. It was the first time I'd been there and I was amazed at what some people were throwing into the skips. One of the operatives was just about to scrap two children's bikes and they looked fairly new. I asked him how much they wanted for the bikes, but he said they weren't allowed to sell anything; everything had to go in the skips, and then head off to landfill. I could not believe it. I was so angry at seeing what they called 'rubbish' right there in front of my eyes that I went straight home and applied to start running a Freecycle group. There was a bit of wrangling in the beginning, deciding on exactly where the boundaries would be, as there were many other groups set up nearby (all great news), but some of

the areas overlapped slightly. It took about nine months to get it up and running, and one year and nine months down the line my group has over 500 members. I give my heartfelt thanks to every one of them for making it what it is.

I've made so many genuine friends here too, and through the group I reckon I've helped liberate enough stuff to fill a full-sized football field, if not more, from the fate of the dreaded rubbish tip and landfill. Visit **www.freecycle.org.uk** for details of your nearest group and sign up.

8. Re-using junk

Do you remember the Wombles? Who could ever forget them? They were a positive part of my youth but, sadly, they dropped off the familiar children's television slot after school.
This next case study spotlights a man who has breathed life back into the concept of 'making good use of the things that we find, things that the everyday folks leave behind.' Some say it's eccentric, others say its a little bit 'out there'. I say its funky, quirky and, more importantly, it's making a difference. Enter, stage left, Mr Junkk.

Peter Martin doesn't keep a tally of his age anymore; he's got far more important things to do getting us all to think re-use. He lives with his wife and twin boys in Ross-on-Wye.

I love making things and really fancied exploring the re-use and repair route to reducing my own personal rubbish. Then I started pondering something larger, more positive and pro-active that would reach lots of other people too. It had reward and incentive at its core and brought in

creativity and fun with the tangible environmental benefit of
less waste being sent to landfill. Our family were already
concocting things in our shed from junk, so it was only
a small step forward to the creation and development of
www.junkk.com.

By pooling the imaginative resources of like-minded
individuals, Junkk comes up with tangible solutions for
anyone who has something they are about to throw out,
but would prefer not to. On the website, we have a funky database and
search system to help people do just that.

All you have to do is visit the website and say, "I have an empty
bottle of 'Brand X gloop,' what can I do with it to spare the bin and the
landfill?" If someone else has figured out a use, then we put you guys
in touch. We've also uploaded product and packaging data into the
system, so you can find new uses for things sitting in your recycling
bin. There's also a great newsletter packed full of tips that you can sign
up to. I've even taken to penning a blog, which is proving quite popular,
if often a little contentious.

There's much more to dig into on the website, which is free to join,
and there's a great deal in development too. I'm looking forward to
realising the ever-evolving potential of the site, in partnership with all
those kind enough to have visited, stayed and who keep coming back,
bringing new friends along too.

The ideas we have on the site are ingenious and amazingly simple.
For example, have you ever thought of putting floor protectors under
your table's legs? A pack of four costs around £2, or you could keep
the caps when your kids have a milkshake and use those instead!
Yes, it's trivial, but it's also four bits of plastic less in landfill, which is
easy to do.

I will not pretend that re-use is going to make a vast difference to
levels of landfill, but it is a great start and it does entice the common
man, woman, child and silver surfer into a mindset that thinks of
refuse as a resource again. It adds, too, to other worthy efforts, such
as recycling.

In every area of our lives we need to exist more sustainably because we live on finite areas of land, surrounded by finite volumes of water, with a finite atmosphere above. The human race, together with our waste, has almost filled up those spaces.

I think radical reductions and mitigations are in order. How, where, by whom and in what order is simply too massive for me as an individual to get my head around, so I've busied myself with personal contributions within our home life and with the creation of Junkk. I'd like to think it may one day be counted as a small acorn that was part of a mighty forest of positive action.

My dream is to influence manufacturers of products to design in a second use for them. I truly believe we can be part of that too; for example, I posted an idea on the website for using old bike inner tubes as door seals and now our local bicycle shop keeps the bin of old ones as a lure for people to pop into his shop, collect them for re-use and perhaps spend a few pounds in the shop while they're there.

I believe that most of the population are ready, willing and able to do their bit. The authorities could set achievable objectives that are honestly explained and clearly communicated, preferably with some form of end benefit in place to encourage uptake.

Junkk is run by me, my wife and a small team of dedicated volunteers. We get a lot of nice mentions and even the odd award from time to time. You can't eat them, of course, but it's always nice to feel appreciated. It's a worthwhile project to have instigated and be part of and, no matter what, it is a legacy I'll always be proud of.

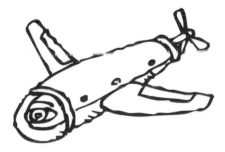

Star struck ...
celebrity questions
and answers ...

Brigit Strawbridge

Brigit is a writer and broadcaster on sustainable living and creator of The Big Green Idea.

Q: You first came to our screens with the popular television series, It's Not Easy Being Green, *and I know you are busy promoting* The Big Green Idea *in villages and towns up and down the country. How easy is it to reduce the personal rubbish you generate when you are on the road, and do communities have a part to play in encouraging their citizens to recycle?*

A: I travel a lot on trains at the moment and find it very frustrating when I want a cup of tea from the buffet. It usually comes in a 'non-recyclable' plastic-coated cup that has to be put in a British Rail paper bag (health and safety!), and the milk comes in one of those tiny plastic containers instead of being poured from a jug.

I've kept a paper bag to re-use and I bring a little milk with me in a small re-sealable tub. I have a little thermos flask for day trips, so avoid using the buffet service entirely, but on overnights I can't take the extra weight and space of the flask in my rucksack.

What I hate most is seeing everyone's rubbish, most of it paper and therefore easily recyclable, being swept into a giant black bin bag at the end of the train journey; it looks like it's all going to landfill. Surely they could sort out a simple recycling system, or encourage people to take their rubbish home with them to recycle there.

Communities have a huge role to play; they're made up of individuals who have the potential to become greater than the sum of their parts. The powers-that-be seem to be oblivious to the seriousness of the problems we face, so it's up to communities to take responsibility. The transition movement is a great example of how communities can pull together to make a difference.

Q: What's your trickiest item to recycle and how do you do it?

A: The trickiest item to recycle is plastic wrapping from food. Now, I keep it if it's clean and plan to use it to stuff outdoor cushions and beanbags, or at the very least shred it and use it again as protective packaging when I send breakable things through the post

For details of *The Big Green Idea*, visit **www.thebiggreenidea.co.uk.**

John Naish

John is a journalist and the author of *Enough*, a practical guide to future-proof consumption.

Q: Your best-selling book Enough *explored some of the many ways consumers can break free from the world of 'more'. How can we help people make the connection between over-abundant consumption and an apparent disregard for the waste and rubbish it produces?*

A: I think it's about encouraging a culture where we give much more value to what we already have. Fashion and consumerism tell us that once we own a thing, it's bound to be 'so over' already and the thing we should value highest is the next consumer item being dangled before us. It's a system that produces goods that aren't made to last (because we aren't meant to value them), and this makes them easy to chuck away. Instead, we should be demanding goods that we can value for their durability and truly good design, and we will then want to hang on to them.

We need to revive the idea of having sentimental long-term relationships with the things that we own. They become part of our life-story, rather than being short-termist consumer items that make us yearn to belong to some superficial brand's artificial life-story.

Q: What's your trickiest item to recycle and how do you do it?

A: It's sump oil from my motorcycles. It's a dirty habit and one of which I'm not too proud. There's no particularly wonderful use for my nasty old hydrocarbons (at least, as far as I'm aware). As it happens, my brother has an oil-burning heater that can swallow the stuff, but it's hardly an ideal solution.

John's book *Enough* is published by Hodder and Stoughton and is available at all good book shops. John is also founder of The Landfill Prize, see **www.enoughness.co.uk** for details.

Wayne and Gerardine Hemingway

Wayne and Gerardine are the designers of award-winning Red or Dead fashion design and now run Hemingway Design.

Q: You both speak out about the effect that architecture and product designs can have on the global landscape. Should new housing projects be incorporating eco-designs to help us reduce our rubbish and lighten our impact on the environment and, if so, where should they begin?

A: Of course they should, and there is increasing legislation in place — The Code For Sustainable Homes — but there is such a long way to go to reduce rubbish and increase recycling. On the whole, we haven't grasped this in the UK. We continue to allow house builders to build identikit rabbit hutches that research regularly shows are largely unloved by a population in the grip of a major housing shortage. As in the 1960s and 1970s, we are building the slums of the future that will be pulled down in 25–30 years.

I am confident that the landscape and house design will suffer because of the rush towards scoring eco-points to obtain planning permission. What can be more unsustainable than building and then pulling down within thirty years? Surely all the renewables won't offset this wasteful carbon footprint.

Communal recycling points, like we have at our development, The Staiths South Bank in Gateshead, encourage recycling, get people out into the streets, can aid sociability, can impact positively on recycling by 'keeping up with the Joneses', and reduce the need for streets that can handle big bin lorries, allowing home-zone style environments. However, housing developments can go much further. Green space could and should be designed to encourage more grow-your-own and composting. We have to tackle the carbon issues, but not at the expense of liveability, or we really will be kicking ourselves in the shin.

Q: What is your trickiest item to recycle and how do you do it?

A: I suppose it's all the cellophane and plastic packaging that so much food comes in. If it seems like the same material as plastic bottles, we add it to them. I think we should also get more imaginative. I like doing thrifty things, such as making dog ends of soap into larger bars by being a creative soap sculpturer.

For further details of Wayne and Gerardine's projects, and to see the eye-catching 'Butt Butt', visit **www.shackup.co.uk** and **www.hemingwaydesign.co.uk**.

Kim Wilde

Kim, an '80s pop star and daughter of Marty Wilde, is now a celebrity gardener.

Q: Nine out of ten homes in America have a waste disposal unit for food waste. Composting is the preferred method in the UK for food waste disposal and it can reduce our bins by around 20%. Why is it important for us to compost our organic waste and how can we get started?

A: Waste disposal units are not widely used in the UK. This method sends valuable nutrients into the sewage system. Compost, by contrast, puts goodness back into the earth while allowing you to reduce the amount of waste you put in your bin. If you don't have a garden or you live in an apartment with no space for a wormery, contact your local authority to find out if there is a food waste collection or disposal facility nearby.

Recycling and being economically efficient is vital for the future of our planet; not recycling can no longer be an option in our daily lives. By taking simple steps to become greener in your kitchen you will help make a difference without too much effort or cost. I'm a busy working mum and have run an environmentally friendly household for years. I'm working with Magnet Kitchens on a project to promote eco-solutions for households. They are working to encourage more homes to make recycling part of their routine and are developing their first kitchen entirely from renewable resources. Magnet are donating some of their profits to The Carbon Trust (see **www.smartplanet.com**).

Q: What's your trickiest item to recycle and how do you do it?

A: To be honest, I don't have much trouble recycling. I bring as much fresh produce into the house as I can, so I don't have too much excess packaging and waste to dispose of. If you change your shopping habits, you can do it. Also, anything that says it's recyclable is exactly that! It's just a question of finding out who collects it, or where to deposit it for reprocessing.

For further information on Kim, visit **www.kimwilde.com**. To find her range of eco-kitchens, please visit **www.magnet.co.uk**.

Joanna Yarrow
Presenter on GMTV and Green Lifestyle Expert

Q: You are founder of the sustainability company Beyond Green *and an accomplished author and television presenter. Do you think we need to change our attitudes and shopping habits and, in an age of 'loving the cheap', how can we steer people towards more expensive, longer lasting and sustainable products?*

A: We've come to associate more with good — we buy things to treat ourselves, cheer ourselves up or affirm our place in life. Rather than seeing more and low cost as king, we need to emphasize the fact that 'cheap' has a cost — to the environment, to producers, and to us. One way or another we pay to throw!

We need to learn to live with less, better, stuff. The growth in ethical products is exciting, it's great to be able to buy sustainable alternatives. But we need to consume less. We need to make choices based on quality, durability, adaptability and beauty. To only buy things we love, cherish and need, and to keep — not throw in a hole in the ground!

Q: What's your trickiest item to recycle and how do you do it?

A: Clothes — I love them! I enforce a strict hierarchy: repair, recondition, swap or sell; take to a charity shop; recycle those that are beyond wearing. I try to buy classic, well-made timeless pieces that last, complemented by items made from recycled materials.

Carl Honoré

Carl is a Canadian journalist who wrote the best-selling book *In Praise of Slow: how a worldwide movement is challenging the cult of speed.*

Q: Your best-selling book, In Praise of Slow, *and your latest title,* Under Pressure, *both compellingly engage the reader to analyse if, and why, they live such busy lives. Do you feel it is important for us to slow down our pace before we can really get to grips with the problems we're creating with our piles of rubbish, and what's your best advice for anyone who's almost too busy to read this book?*

A: Slowing down is definitely the first step to tackling our waste problem. Putting on the brakes implies consuming less, which means wasting less. However, it also gives us the time and space to ask the deeper questions about why we consume so frenetically in the first place, and what our true needs really are. To someone who feels too busy to read this book, I would put to them a simple question: 'How much of what you buy today will you remember on your deathbed?'

Q: What's your trickiest item to recycle and how do you do it?

A: Cardboard boxes covered in tape, stickers and plastic wrapping. I get loads of these and they drive me nuts. Recycling them involves complex origami and struggling to unstick the perma-stuck. My solution is to enlist the children. Either they keep the boxes and turn them into houses, spaceships, robots or whatever, or we have some fun doing the origami and the unsticking together. Either way, the boxes end up recycled.

For further information on Carl and his books, and talks on living life at a slower pace, visit **www.carlhonore.com**. For a useful guide to slow holidays read *Go Slow England*, published by Alastair Sawday Publishing.

Penney Poyzer

Penney is a journalist, author, speaker, trainer
and broadcaster and has worked in the green sector
for over ten years.

*Q: You interviewed people up and down the country for your
groundbreaking television series,* No Waste Like Home. *It must have
been a heart-warming moment when they finally realised how much of
a problem we have in our landfill sites. Why do you think we have got to
stop dilly-dallying around with this topic and take personal responsibility
for our rubbish?*

A: It is one of those questions where you think, 'It is obvious to me,
why on earth doesn't everyone see it that way?' It's very frustrating, but
I think waste has become very ingrained. Food waste regularly hits the
news, with around a third of what we buy going in the bin.

I think it is down to a couple of issues. It is very easy to make waste
because of the way we shop. Supermarkets don't encourage people to
be sensible; the more they sell, the bigger the profits. Recent food price
increases have added £15 to the average weekly shop and will, I hope,
encourage more people to be sensible.

Government agencies need to work closely with supermarkets to
encourage good use of leftovers. The first supermarket that does
encourage common-sense shopping could score a massive PR coup.
I would love to see M&S start up a campaign with a high-profile chef,
and a book, on how to use up leftovers and how to shop wisely and
avoid waste.

In 1957 we spent a third of the available household budget on food;
today it is a mere 15%. Food comes around twentieth on the list of
household expenditure. We don't value food, so we sling huge amounts
of food and packaging in the bin.

People want to give their families choice when it comes to food,
and it is common to find a mum dishing up two or three different

dishes at mealtimes to cope with fussy eaters. I think it needs a massive national campaign to target mums; they do the shopping and, while it is uncomfortable to think about it, they are largely responsible for the waste!

Q: What's your trickiest item to recycle and how do you do it?

A: Interesting question. I'm lucky because where I live our council recycles most things apart from glass, which we take to our local bottle bank. Plastic sleeves are the trickiest; no one recycles it because there is no market. I don't buy pre-packaged food so it is not an issue for me. I do manage to recycle pretty much everything!

Penney is the undisputed Queen of Green. Her book, *No Waste Like Home*, which accompanied her BBC television series, is available in all good book shops. It is published by Virgin Books.

Janey Lee Grace

Janey Lee Grace is a singer, author, television presenter and radio disc jockey.

Q: Your work as a broadcaster and author on holistic living is inspiring, and it's encouraging to see broadcast media helping to get the sustainable living message out there. But how much effort are they making to reduce their rubbish and to encourage others to 'tread lightly' behind the scenes?

A: I can't comment on what the broadcast media do behind the scenes, I'm usually on the frontline! I think most offices and studios have greened up a bit and at least have recycling facilities in place, but I'm sure there's a lot more that can be done.

Q: What's your trickiest item to recycle and how do you do it?

A: CDs! Working in broadcasting I get given lots of CDs for review, which is nice, but they're more redundant as people use iPods etc. Any collectable ones I sell at car boot sales and give the money to charity and the rest we turn into art statements. The kids have made a wonderful outdoor art design from CDs; they look wonderful when they catch the sunlight.

You can find full details of Janey and her best-selling *Imperfectly Natural* series of books — available at all good book stores and directly on her website — by visiting **www.imperfectlynatural.com**.

Abi Roberts

Described as the British Bette Midler, Abi Roberts is a genre-defying, multi-talented stand-up, music and theatre performer.

Q: I know you are passionate about environmental issues and as a top entertainer on the comedy circuit, it must be difficult to keep on top of recycling and rubbish reduction when you're on the road. Do you have any tips for people who live out of a suitcase for long periods of time?

A: I do a lot of touring and travelling and do tend to live out of a suitcase. I find the best way to reduce rubbish is to use any plastic bags I have for laundry or to put my stilettos in (I don't do flats, sweetie!). Also, I hand-wash my laundry whenever I can; there's plenty of draping space over hotel baths. Don't pay five quid for a pair of socks to be washed by the hotel laundry.

Also, if you buy newspapers or magazines check that the hotel or motel (can you tell I'm in the US of A?) has a recycling bin outside. If not, then keep a bag of papers until you find one.

If you are buying something and the shop assistant says, 'Do you want a plastic bag?' just smile and say, 'Do I look like the kind of person who collects plastic bags?' Might be best to leave it a few days before going in again.

Diva's Top Tip: Stuff old magazines into your Manolos (shoes) to keep them in shape; that's what Sarah Jessica Parker does apparently.

Q: Do you do any green comedy and how does it go down?

A: I have started to do more 'green' comedy. Comedians are the greenest people on earth, as most of them recycle their material — over and over again, particularly the well-known ones; hush my mouth. One of my favourite jokes is one I used recently: 'What's the difference between a hedgehog and a 4×4? With a hedgehog, the pricks are on the outside!' I tried a bit about how we should stop moaning about closing down the mines in the 1980s, as it actually reduced carbon emissions. Note to self: don't try that again in Newcastle.

Q: What's your trickiest item to recycle and how do you do it?

A: That's a tricky one (she says, stating the obvious). I mean you can recycle most things now. I understand we girls can even recycle our worn-out rabbit vibrators, so I guess the one thing I would have a problem with would be the batteries! But men are by far the trickiest to recycle. They're just not the same the second time round. We are all like students sharing a rented house — except Mother Earth is the landlady and she's not worried about pets or gentlemen callers. I don't want to be evicted before my finals.

For further information on Abi, including tour dates, see **www.abiroberts.com** and do visit the eco-friendly website of her ever-inspirational mum, **www.zacharrys.co.uk**.

Solutions

We've been through the house from top to bottom, I've presented you with simple options for minimising the rubbish you create, and I hope it has inspired you to look at waste in a different light. However, if we are going to make significant changes to the trash we generate as individuals and on a community and global platform, we need to change our habits on a grand scale.

The benefits of positively embracing living with less are plentiful — more money in our pockets, a reduction in overburdened landfill sites, minimised impact on our climate and environments — but it runs far deeper than that. If we hold more respect for the things we own and really consider the implications and consequences of continuing to live a throwaway lifestyle, we will be well on the way to helping close the loop in the materials economy and creating a better future for our children and future generations.

We would do well to remember the make-do-and-mend mentality of the war years. There have even been mutterings at Westminster, encouraging us to grow a few things to eat; I wonder if the Prime Minister will set us a good example by starting a vegetable plot in the gardens at Number 10? If he does, we might see an increase in the number of community allotments being offered and, maybe, comprehensive sustainable-living lessons in the national curriculum to help the children find their green groove.

The simple fact remains, whatever we're told by whoever is in power, that the buck really does start and end with us as individuals, and we have the power to effect positive change. Start by creating your own 'Do I really need this?' checklist to take out when you go shopping, and try to eliminate one superfluous item a week from every shopping trip.

Share your new-found enthusiasm for waste reduction in your street, your school and your workplace. Encourage simple schemes to be put in place and shout at the retailers to bring about change in your community; the email templates in this book will help get you started.

Look at the successful Zero Waste Week campaigns like the one in St Edmundsbury and others that have run in this country and overseas. If this initiative took place throughout the UK, we would see healthy competition everywhere, with councils fighting to be top green dog. We would all benefit from local authorities walking the walk, not just talking the talk.

We need to start doing things in a different way. The disposable era must come to an end before it consumes us and our planet. Make re-usable, refillable and rechargeable your new shopping mantras, and think about the size and longevity of what you're putting in your trolley. Minimise waste wherever you can and use products that do more than one thing.

Recycle and buy recycled. Take a look at what imaginative designers and manufacturers are making with recycled materials.

Consider the transport miles for food and the everyday items you consume, and make one small and simple change to what you do each week. In fact, think about the transport miles for everything you buy.

Start composting; it's so easy and there are options for us all wherever we live. It will reduce the immense pressure on our disposal facilities and landfill sites and give us something to feel good about.

Nudge the responsibility for recycling back into the hands of the retailers. Resist packaging, say 'no' to mass consumerism, put more value on the things you have, and pass this book onto a friend.

Take on board some of the free advice on the websites of WRAP and Recycle Now. Both help individuals and corporations to make changes. Get involved with the annual Recycle Week. In 2008, to mark the event, accomplished sculptor Robert Bradford helped launch Recycle Week by exhibiting a series of inspirational pieces, including a six-metre high sculpture of Big Ben made from drinks cans. Visit **www.robertbradford. co.uk** to view them all. Consider doing something in your immediate vicinity to show your community's commitment to change.

Perhaps the best advice I can offer is to encourage you to take up the challenge of rubbish reduction with a positive attitude, and just do as much as you can.

Advice and information

Baby Milk Action
Baby Milk Action aims to save lives and reduce the suffering caused by inappropriate infant feeding.
www.babymilkaction.org
Baby Milk Action, 34 Trumpington Street, Cambridge, CB2 1QY, UK
Tel: 01223 464420

Battery Hen Welfare Trust
Working to inspire a free-range future.
www.bhwt.org.uk
Tel: 01769 580310

Belfast City Council
Waste reduction for Belfast.
www.belfastcity.gov.uk/Recycle/
Belfast City Council, Adelaide Exchange, 24–26 Adelaide Street, Belfast BT2 8GD
Tel: 028 9032 0202

Big Barn
Find farmers' markets, vegetable and fruit box schemes, artisan bakers and pick-your-own sites.
www.bigbarn.co.uk
Tel: 01234 871005

Big Green Switch
Guide to greening up your lifestyle by making small, simple, everyday changes.
www.biggreenswitch.co.uk

Brita Water Filters
All cartridges are recyclable.
www.brita.net
Brita Water Filter Systems Ltd, Brita House, 9 Granville Way, Bicester OX26 4JT
Tel: 0844 742 4900

British Aerosol Manufacturers' Association
Support and advice services to members.
www.BAMA.co.uk
BAMA, King's Buildings, Smith Square, London SW1P 3JJ
Tel: 020 7828 5111

British Plastics Federation (BPF)
Trade association of the UK Plastics Industry.
www.bpf.co.uk
The British Plastics Federation, 6 Bath Place, Rivington Street, London EC2A 3JE
Tel: 020 7457 5000

Car Boot Junction
Informative directory of car boot sales in the UK.
www.carbootjunction.com
Email: enquiries@carbootjunction.com

Charity Shops
Supporting charities that run shops where you can buy, donate and volunteer.
www.charityshops.org.uk
Charity Shops, 5th Floor, Central House, Upper Woburn Place, London WC1H 0AE.
Tel: 020 7255 4470

Chicken Out Campaign
Striving to improve conditions for our poultry.
www.chickenout.tv
Email: campaign@chickenout.tv

Combined Heat and Power
Sustainable power solutions, including solar, wind and biomass throughout the UK.
www.microgeneration.co.uk
Tel: 0845 434 8084

Community Composting Network
Promoting home composting.
www.communitycompost.org
Community Composting Network, 67 Alexandra Road, Sheffield S2 3EE
Tel: 0114 2580483 or 0114 2553720

Community RePaint
Collects re-usable paint for redistribution to community and voluntary groups, charities and individuals.
www.communityrepaint.org.uk
Tel: 0113 200 3959
Community RePaint, c/o Resource Futures, 3rd Floor, Munro House, Duke Street, Leeds LS9 8AG

Community Recycling Network (CRN) UK
Promoting community-based waste management
as a practical and effective way of tackling the
UK's growing waste problem.
www.crn.org.uk
The Community Recycling Network UK,
57 Prince Street, Bristol BS1 4QH
Tel: 0117 942 0142

Compost Association
Sustainable management of
biodegradable resources.
www.compost.org.uk
Tel: 0870 160 3270
The Composting Association, 3 Burystead Place,
Wellingborough, Northamptonshire NN8 1AH

DEFRA
Dept for Environment, Food and Rural Affairs
www.defra.gov.uk/environment/statistics
DEFRA Customer Contact Unit,
Eastbury House, 30–34 Albert Embankment,
London SE1 7TL
Tel: 08459 335577

Direct Marketing Association (UK) Ltd
Contact them to reduce the delivery of
unaddressed mail.
www.dma.org.uk
Your Choice Preference Scheme, Direct
Marketing Association (UK) Ltd, DMA House,
70 Margaret Street, London W1W 8SS
Tel: 020 7291 3300

Encams
Keep Britain Tidy campaign and other campaigns.
www.encams.org
Tel: 01942 612617

Energizer Ltd
Recycle your old batteries.
Energizer Ltd, Recycling, Freepost LOL2311,
Dunstable, Bedfordshire LU5 4YY

Envocare
A central source of data on environmental issues.
www.envocare.co.uk

Environment Agency
Covers all aspects of environmental concerns
and publishes an online magazine called
Your Environment.
www.environment-agency.gov.uk
Environment Agency, National Customer Contact
Centre, PO Box 544, Rotherham S60 1BY
Tel: 08708 506 506 (Mon–Fri 8–6)

Freecycle
A worldwide network of online groups giving
stuff away to those who need or want it.
www.freecycle.org.uk

Free2Collect
A non-profit site that is promoting a nationwide
effort to recycle unwanted items.
www.free2collect.co.uk

Friends of the Earth
www.foe.co.uk
Friends of the Earth, 26–28 Underwood Street,
London N1 7JQ
Tel: 020 7490 1555

Furniture Re-use Network
Supports, assists and develops charitable re-use.
www.frn.org.uk
FRN Office, 48–54 West Street, St Philips,
Bristol BS2 0BL
Tel: 0117 954 3571

Green Choices
Direct information on green lifestyle alternatives.
www.greenchoices.org.uk

Green Steps
Information, advice, products and services.
www.greensteps.co.uk
Green Steps Limited, Austral Farm, Burnham
Road, Chelmsford, Essex CM3 6DP
Tel: 0845 416 1671 or 07970 190602

Greenpeace
www.greenpeace.org
Greenpeace, Canonbury Villas, London N1 2PN
Tel: 020 7865 8100

Help the Aged
International charity fighting to free older people from poverty, isolation and neglect.
www.helptheaged.org.uk

INCPEN
Aims to ensure that policy on packaging makes a positive contribution to sustainability.
www.incpen.org
INCPEN, Soane Point, 0–8 Market Place, Reading, Berkshire RG1 2EG
Tel: 0118 925 5991

InterNational Downshifting Week
Ideas on how to slow down and green up your life.
www.downshiftingweek.com

Lithium Powered Tools
For a range of DIY tools using recyclable batteries.
www.bosch.co.uk

Local Exchange Trading Scheme (LETS)
Find your local network.
www.letslinkuk.org

Love Food Hate Waste
Information and a brilliant campaign to reduce food waste.
www.lovefoodhatewaste.com
Tel: 0808 1002040

Love Honey
Get involved in the Rabbit Amnesty!
www.lovehoney.co.uk
Tel: 0800 915 6635

Love Libraries
Find out about their latest campaigns and great things that are happening in libraries all around the country.
www.lovelibraries.co.uk
Tel: 020 7273 1432

Magdalen Project
An eco-educational centre in Somerset that runs courses in rural crafts.
www.themagdalenproject.org.uk

Mailing Preference Service (MPS)
A free service to enable consumers to remove their names and home addresses from lists used by the industry.
www.mpsonline.org.uk
Mailing Preference Service, DMA House, 70 Margaret Street,
London W1W 8SS
MPS Registration line: 0845 703 4599

National Association for Children of Alcoholics (NACOA)
Supporting children of all ages who have one or two alcoholic parents.
www.nacoa.org.uk
NACOA PO Box 64, Fishponds,
Bristol BS16 2UH
Tel: 0117 924 8005; Helpline: 0800 358 3456

National Household Hazardous Waste Forum (NHHWF)
Advice on the management of household hazardous waste, including collection, recycling and disposal.
www.nhhwf.org.uk
NHHWF, Third Floor, Munro House, Duke Street, Leeds LS9 8AG

Oil Care Campaign
Provide guidance and facilities for the safe disposal and management of oil.
www.oilbankline.org.uk
Tel: 08708 506 506

Ollies World
Lots of initiatives for children to become involved in environmental fun activities.
www.ollierecycles.com
Sustain Ability International Pty Ltd,
P.O. Box 75, Camberwell, Victoria 3124,
Australia

PaperOnWeb
Pulp and paper resource and information site.
www.paperonweb.com
Email: harigoyal@yahoo.com

Polyprint Mailing Films
Promotes the recycling of polythene products.
www.polyprint.co.uk/recycling.html
Polyprint Mailing Films Ltd, Mackintosh Road,
Rackheath Ind. Est., Rackheath,
Norwich NR13 6LJ
Tel: 01606 721807

Real Nappy Campaign
Find a real nappy consultant in your area.
www.realnappycampaign.com
Tel: 0845 850 0606

Reclaim-it
Send mobile phones and printer cartridges.
www.reclaim-it.com
Tel: 0870 7744288

Recoup
Works to maximise efficient plastics recycling.
www.recoup.org
1 Metro Centre, Welbeck Way, Woodston,
Peterborough PE2 7UH
Tel: 01733 390021

Recycle for London
Making it easier to recycle more.
www.recycleforlondon.com
Recycle for London, Greater London Authority,
City Hall, The Queen's Walk, London SE1 2AA
Helpline: 0845 331 3131

Recycle for Wales
Information to the public about managing
materials and resources more sustainably, and
reducing waste.
www.wasteawarenesswales.org.uk
Waste Awareness Wales, Local Government
House, Drake Walk, Cardiff CF10 4LG

Recycle Now
Ideas, advice and inspiration to help you recycle.
www.recyclenow.com
Tel: 0845 331 3131

Recycled Products Guide
Provides a national, comprehensive database of

products made from recycled materials.
www.recycledproducts.org.uk
Email: rpg@wrap.org.uk
Contact by email only.

Recycling Consortium
Managing many influential recycling projects.
www.recyclingconsortium.org.uk
The Recycling Consortium, CREATE Centre,
Smeaton Road, Bristol BS1 6XN
Tel: 0117 930 4355

Royal Mail Door-to-Door Opt Outs
To opt out of receiving door-to-door mail items,
send your name and address to:
www.mydm.co.uk
Freepost RRBT-ZBXB-TTTS, Royal Mail Door-to-
Door Opt Outs, Kingsmead House, Oxpens Road,
Oxford OX1 1RX
Tel: 0845 703 4599.

Royal Society for the Prevention of Accidents
(RoSPA)
www.rospa.com

Royal Society for the Protection of Birds (RSPB)
www.rspb.org.uk

Rubbish Diet
Blog by Karen Cannard.
www.therubbishdiet.co.uk

Salvo
Directory to reclamation yards in the UK and
worldwide.
www.salvo.co.uk
Tel: 01225 422300

Smart Planet
News, reviews and tools to make it easier to
create a greener world.
www.smartplanet.com
Tel: 020 7021 1000

Soles United
Recycling footwear donation scheme
www.solesunited.com

St Edmundsbury Council
The successful 'Zero Waste Week' case study.
www.stedmundsbury.gov.uk/zerowaste

Steel Can Recycling Information Bureau
All you need to know about recycling cans.
www.scrib.org
Steel Can Recycling Information Bureau, Trostre
Works, Llanelli, Camarthenshire SA14 9SD
Tel: 01554 712632

Sustainable House Information
www.alanmontague.com
www.thetimberframe.co.uk
www.energysavingtrust.org.uk

Textiles On Line
Lots of great information and interactive
elements for individuals and schools.
www.e4s.org.uk/textilesonline

Sort It
Information and practical advice on a range of
local and national recycling and waste reduction
facilities and services in Scotland.
www.sort-it.org.uk
Tel: 01786 468 245

Store
Great storage and recycling receptacle ideas.
www.aplaceforeverything.co.uk

Transition Towns
Transition initiative to help with life after oil.
www.transitiontowns.org

Waste Aware Scotland
Advice on how and why to reduce, re-use and
recycle your household waste.
www.wasteawarescotland.org.uk
Scottish Waste Awareness Group, Enquiries,
Islay House, Livilands Lane, Stirling FK8 2BG
Tel: 01786 468 248

Waste Watch
Action on waste reduction, re-use and recycling.
www.wastewatch.org.uk

Waste Watch, 56–64 Leonard Street,
London EC2A 4LT
Tel: 020 7549 0300

Waste Online
Information on waste minimisation and
waste reduction.
www.wasteonline.org.uk

WEEE Directive
Waste electrical and electronic equipment
recycling scheme.
www.weeecare.com

Woodland Trust
Charity for trees and woodlands throughout the UK.
www.woodland-trust.org.uk
The Woodland Trust (England), Autumn Park,
Dysart Road, Grantham, Lincolnshire NG31 6LL
Tel: 01476 581111
Woodland Trust (Scotland), South Inch Business
Centre, Shore Road, Perth PH2 8BW
Tel: 01738 635829

WRAP (Waste & Resources Action Programme)
Encourages and enables businesses and
consumers to be more efficient in their use of
materials, and to recycle.
www.wrap.org.uk
Helpline: 0808 100 2040
Waste & Resources Action Programme, The Old
Academy, 21 Horse Fair, Banbury,
Oxon OX16 0AH

WRAP for Northern Island
www.wrap.org.uk/nations_and_english_regions/
northern_ireland/index.html
Tel: 0870 3515180

Yellow Pages
To cancel your copy call 0800 671444.
Visit www.yell.com for an online directory.

Zero Waste Alliance UK
A cleaner and healthier way to deal with waste.
www.zwallianceuk.org
Email: val@zwallianceuk.org

Eco products

Baby Kind
Promoting the use of cloth nappies.
www.babykind.co.uk

Bay House Aromatics
Essential oils for many household cleaning tasks
and personal use.
www.bay-house.co.uk

Bio-D
A UK-based range of eco-friendly cleaning products.
www.biodegradable.biz

Bishopston Trading Company
A pioneer fairtrade clothing company.
www.bishopstontrading.co.uk

Botanicals
Plant-based and organic products with no
artificial additives, colourings, harsh chemicals or
preservatives.
www.botanicals.co.uk

Buy Recycled
Supplies cartridges for most inkjet and laser
printers.
www.buyrecycled.co.uk

Charity Gifts
The official online charity shopping portal.
www.charitygifts.com

Eco-Craft
Recycled paper and card, envelopes and card blanks.
www.eco-craft.co.uk

Ecomundi
A great range of biodegradable party wear.
www.ecomundi.co.uk

Eco Wood Man
Hand-made gifts using sustainably sourced and
reclaimed wood.
www.ecowoodman.co.uk

Eco-Wool
Insulation made from 80% recycled plastic
bottles.
www.eco-wool.co.uk

Felicity Hat
Hat hire and accessories company.
www.felicity.co.uk

Good Energy Shop
Torches, energy-efficient light fittings, renewable
energy-charging devices and much more besides.
www.goodenergyshop.co.uk

Green Shop, The
Sustainable and low-impact products for homes.
www.greenshop.co.uk

Handbag Hire
Designer bags.
www.handbaghirehq.co.uk

Home Recycling
Recycling bins, crushers and other
recycling tools.
www.homerecycling.co.uk

Natural Collection
Award-winning online green shop.
www.naturalcollection.com

Neal's Yard Remedies
Organic skin and body care and natural remedies.
www.nealsyardremedies.com

Nigel's Eco Store
For wooden cutlery and magnetic eco-balls.
www.nigelsecostore.com

Nouvelle
Producers of recycled toilet tissue.
www.nouvellerecycling.co.uk

Original Organics
For all your wormery and composting needs.
www.originalorganics.co.uk

Pet Planet
Suppliers of Good Boy dog loo.
www.petplanet.co.uk

Rapanui
A green stockist of funky surf and casual wear made from ethical and eco-friendly fabrics.
www.rapanuiclothing.com

Riverford Organic Home Delivery
Organic fruit and vegetable box delivery service.
www.riverford.co.uk

Seasalt
Hardwearing and fashionable clothes available online and in stores that feature a range of organic items.
www.seasaltcornwall.co.uk

Second Nature
Insulation (Thermafleece) made using wool from sheep grazing upland areas of the UK.
www.secondnatureuk.com

Vessel
Offering a great range of low-energy, eco-lighting products.
www.vessel.com

Water Butts
Funky rainwater collection butts.
www.waterbutts.com

Who Cares?
Ethical and fairtrade supermarket.
www.whocares.gb.com

Bibliography

A Women's Guide to Saving the World, Karen Eberhardt-Shelton, 2008

Bread: The Definitive Guide, Sara Lewis, Hamlyn, 2007

Car Sick: Solutions for our Car Addicted Culture, Lynn Sloman, Green Books, 2006

First Time Gardener, Kim Wilde, Collins, 2008

Green Guide to Weddings, Jen Marsden, Green Guides, 2008

Hens in the Garden, Eggs in the Kitchen, Charlotte Popescu, Cavalier, 2003

How to Store your Garden Produce, Piers Warren, Green Books, 2003

Imperfectly Natural Woman, Janey Lee Grace, Crown House, 2006

Natural Stain Remover, Angela Martin, Apple, reprinted 2006

No Waste Like Home, Penney Poyzer, Virgin, 2005

One Planet Living, Pooran Desai & Paul King, Alastair Sawday Publishing, 2006

Save Cash & Save the Planet, Friends of the Earth, Collins, 2005

The Abel & Cole Cookbook, Keith Abel, Collins, 2008

The Converging World, John Pontin, Piatkus, 2008

The Good Shopping Guide, Ethical Marketing Group, annually

The Organic Directory, Clive Litchfield, Green Books, annually

The Transition Handbook, Rob Hopkins, Green Books, 2008

Using the Plot – Tales of an Allotment Chef, Paul Merrett, Collins, 2008

Your Planet Needs You, Jon Symes & Phil Turner, 2006

About the author

Tracey Smith was born in London's East End in 1966 and raised on various council estates, usually without gardens; she'd never cultivated anything more than mould on old cheese. Green-fingered inexperience gave her a shock following an extreme downshift in 2002, when Tracey and husband Ray decided to grow their own, rear their own, and sometimes even kill their own food. At that point, Tracey focused her journalism on sustainable living, unravelling the layers and levels of downshifting, unlocking the secrets of simpler, happier living, and helping people pull back from a consumer-driven existence while being kinder to the environment. She is now a broadcaster on the subject, and volunteers on a show on Apple AM, her local hospital radio station. Tracey writes for several 'green' magazines including the Environment Agency magazine, *Your Environment.*

Smith proffers a nugget of good advice for life generally. She says, 'Ditch the guilt for what you are not doing and start feeling good about what you are doing! Lean towards the green, embrace a bit of decluttering of the mind and the house, and make a fine start today by getting to grips with your rubbish!'

Author acknowledgements

My grateful thanks go to the guys at Recycle Now, WRAP and the Love Food Hate Waste campaign, all working so hard to get people on the right, green track and on a note closer to home, Ray, thank you for putting up with the most disorganised but loving wife.

Environmental award winner 2008
"With green issues currently at the forefront of publishers' minds, **Alastair Sawday Publishing** was singled out in this category as a model for all independents to follow. Its efforts to reduce waste in its office and supply chain have reduced the company's environmental impact, and it works closely with staff to identify more areas of improvement."
Independent Publishers Guild